C0-AYM-004

MADELEINE CLARK WALLACE LIBRARY
WHEATON   COLLEGE
NORTON, MASSACHUSETTS

Hewitt and Ann Fletcher '39
Library Fund

Challenges of Information Technology Education in the 21st Century

Eli Cohen
Leon Kozminski Academy of
Entrepreneurship & Management, Poland
and
Informing Science Institute, USA

 Idea Group
Publishing

 Information Science
Publishing

Hershey • London • Melbourne • Singapore • Beijing

Acquisition Editor:	Mehdi Khosrowpour
Managing Editor:	Jan Travers
Development Editor:	Michele Rossi
Copy Editor:	Maria Boyer
Typesetter:	LeAnn Whitcomb
Cover Design:	Tedi Wingard
Printed at:	Integrated Book Technology

Published in the United States of America by
   Idea Group Publishing
   1331 E. Chocolate Avenue
   Hershey PA 17033-1117
   Tel: 717-533-8845
   Fax: 717-533-8661
   E-mail: cust@idea-group.com
   Web site: http://www.idea-group.com

and in the United Kingdom by
   Idea Group Publishing
   3 Henrietta Street
   Covent Garden
   London WC2E 8LU
   Tel: 44 20 7240 0856
   Fax: 44 20 7379 3313
   Web site: http://www.eurospan.co.uk

Copyright © 2002 by Idea Group Publishing. All rights reserved. No part of this book may be reproduced in any form or by any means, electronic or mechanical, including photocopying, without written permission from the publisher.

Library of Congress Cataloging-in-Publication Data

Challenges of information technology education in the 21st century / [edited by] Eli Cohen.
     p. cm.
  Includes bibliographical references and index.
  ISBN 1-930708-34-3
  1. Information technology--Study and teaching (Higher) I. Cohen, Eli B.

T58.5 .C42 2001
004'.071--dc21   2001051665

British Cataloguing in Publication Data
A Cataloguing in Publication record for this book is available from the British Library.

NEW from Idea Group Publishing

- **Data Mining: A Heuristic Approach**
  Hussein Aly Abbass, Ruhul Amin Sarker and Charles S. Newton/1-930708-25-4
- **Managing Information Technology in Small Business: Challenges and Solutions**
  Stephen Burgess/1-930708-35-1
- **Managing Web Usage in the Workplace: A Social, Ethical and Legal Perspective**
  Murugan Anandarajan and Claire Simmers/1-930708-18-1
- **Challenges of Information Technology Education in the 21st Century**
  Eli Cohen/1-930708-34-3
- **Social Responsibility in the Information Age: Issues and Controversies**
  Gurpreet Dhillon/1-930708-11-4
- **Database Integrity: Challenges and Solutions**
  Jorge H. Doorn and Laura Rivero/ 1-930708-38-6
- **Managing Virtual Web Organizations in the 21st Century: Issues and Challenges**
  Ulrich Franke/1-930708-24-6
- **Managing Business with Electronic Commerce: Issues and Trends**
  Aryya Gangopadhyay/ 1-930708-12-2
- **Electronic Government: Design, Applications and Management**
  Åke Grönlund/1-930708-19-X
- **Knowledge Media in Healthcare: Opportunities and Challenges**
  Rolf Grutter/ 1-930708-13-0
- **Internet Management Issues: A Global Perspective**
  John D. Haynes/1-930708-21-1
- **Enterprise Resource Planning: Global Opportunities and Challenges**
  Liaquat Hossain, Jon David Patrick and M.A. Rashid/1-930708-36-X
- **The Design and Management of Effective Distance Learning Programs**
  Richard Discenza, Caroline Howard, and Karen Schenk/1-930708-20-3
- **Multirate Systems: Design and Applications**
  Gordana Jovanovic-Dolecek/1-930708-30-0
- **Managing IT/Community Partnerships in the 21st Century**
  Jonathan Lazar/1-930708-33-5
- **Multimedia Networking: Technology, Management and Applications**
  Syed Mahbubur Rahman/ 1-930708-14-9
- **Cases on Worldwide E-Commerce: Theory in Action**
  Mahesh Raisinghani/ 1-930708-27-0
- **Designing Instruction for Technology-Enhanced Learning**
  Patricia L. Rogers/ 1-930708-28-9
- **Heuristic and Optimization for Knowledge Discovery**
  Ruhul Amin Sarker, Hussein Aly Abbass and Charles Newton/1-930708-26-2
- **Distributed Multimedia Databases: Techniques and Applications**
  Timothy K. Shih/1-930708-29-7
- **Neural Networks in Business: Techniques and Applications**
  Kate Smith and Jatinder Gupta/ 1-930708-31-9
- **Information Technology and Collective Obligations: Topics and Debate**
  Robert Skovira/ 1-930708-37-8
- **Managing the Human Side of Information Technology: Challenges and Solutions**
  Edward Szewczak and Coral Snodgrass/1-930708-32-7
- **Cases on Global IT Applications and Management: Successes and Pitfalls**
  Felix B. Tan/1-930708-16-5
- **Enterprise Networking: Multilayer Switching and Applications**
  Vasilis Theoharakis and Dimitrios Serpanos/1-930708-17-3
- **Measuring the Value of Information Technology**
  Han T.M. van der Zee/ 1-930708-08-4
- **Business to Business Electronic Commerce: Challenges and Solutions**
  Merrill Warkentin/1-930708-09-2

Excellent additions to your library!

**Receive the Idea Group Publishing catalog with descriptions of these books by
calling, toll free 1/800-345-4332
or visit the IGP Online Bookstore at: http://www.idea-group.com!**

Challenges of Information Technology Education in the 21st Century

Table of Contents

Preface i
    Eli Cohen, Informing Science Institute, USA

Section I: Examples on How to Teach Specific Topics

Chapter I
Teaching Teamwork in Information Systems 1
    Connie E. Wells, Roosevelt University, USA

Chapter II
The Challenge of Teaching Research Skills to Information
Systems and Technology Students 25
        Beverley G. Hope, Victoria University of Wellington, New Zealand
            and City University of Hong Kong
        Mariam Fergusson, PricewaterhouseCoopers, Australia

Chapter III
Data Modeling: A Vehicle for Teaching Creative Problem
Solving and Critical Appraisal Skills 41
        Claire Atkins, Nelson Marlborough Institute of Technology, New
            Zealand

Section II: Teaching Techniques and Pedagogy

Chapter IV
Towards Establishing the Best Ways to Teach and Learn about IT 57
        Chris Cope, Lorraine Staehr, and Pat Horan
        La Trobe University, Bendigo, Australia

Chapter V
Computer-Supported Learning of Information Systems:
Matching Pedagogy With Technology 85
        Raquel Benbunan-Fich and Leigh Stelzer
        Seton Hall University, USA

Chapter VI
Problem-Based Learning in Information
Systems Analysis and Design 100
    John Bentley and Geoff Sandy, Victoria University, Australia
    Glenn Lowry, United Arab Emirates University

Section III: Impact of the Web on IT Teaching

Chapter VII
Teaching or Technology: Who's Driving the Bandwagon? 125
    Geoffrey C. Mitchell, Victoria University of Wellington, New
      Zealand
    Beverley G. Hope, Victoria University of Wellington, New Zealand
      and City University of Hong Kong, China

Chapter VIII
Delivering Course Material via the Web: An Introduction 146
    Karen S. Nantz and Terry D. Lundgren
    Eastern Illinois University, USA

Chapter IX
Bridging the Industry-University Gap: An Action Research
 Study of a Web-Enabled Course Partnership 166
    Ned Kock, Temple University, USA
    Camille Auspitz and Brad King, Day & Zimmermann, Inc., USA

Section IV: Developing an IT Curriculum

Chapter X
E-Commerce Curriculum Strategies and Implementation Tactics: An In-
Depth Examination of DePaul University's Experience 187
    Linda V. Knight and Susy S. Chan
    DePaul University, USA

Chapter XI
Information Systems Curriculum Development as an
Ecological Process 206
    Arthur Tatnall, Victoria University of Technology, Australia
    Bill Davey, RMIT University, Australia

Chapter XII
Information Technology Model Curricula Analysis 222
      Anthony Scime', State University of New York at Brockport, USA

Chapter XIII
Curriculum Model of the Information Resource Management Association
and the Data Administration Managers Association 240
      Eli Cohen, Leon Kozminski Academy of Entrepreneurship
      and Management, Poland, and Informing Science Institute, USA

About the Authors 261

Index 267

Preface

In the two decades that I have been involved with IT education, the field has not just undergone evolution, it has undergone revolution. Both the degree and the speed of change have been spectacular.

One of the changes has been in what we teach. Twenty years ago, most schools taught on and about mainframe computers. Today you would be hard pressed to even find a mainframe computer at most universities. The emphasis has changed first to minicomputers, and then to microcomputers, and now to web-based networked computing.

Another change has been in how we teach. Certainly the technologies have given students greater access to information technologies. But even more fundamentally, we have learned from our experiences and research into how students learn and how best to teach them. Slowly but surely, IT education has taken its rightful place as an important research-based discipline.

WHAT IS IN THIS BOOK?

Here is what you will find in this book. The 13 chapters are divided into four major sections.

Sections

The book is divided into four major sections, the topics of which move from the particular to the general:
1. Examples on How to Teach Specific Topics
2. Teaching Techniques and Pedagogy
3. Impact of the Web on IT Teaching
4. Developing an IT Curriculum

Section 1: Examples on How to Teach Specific Topics

The first section, "Examples on How to Teach Specific Topics," is composed of three chapters. The first is "Teaching Teamwork in Information Systems" by Connie Wells of Roosevelt University (USA). As Connie points out, knowing how to effectively work in teams is very important for those developing information systems. But do we teach teamwork in our courses, or do we just throw students into groups? Connie shows us how to effectively form, build, manage, and evaluate teams in information systems courses. She argues for explicitly teaching students these concepts in addition to the regular course content. This chapter also addresses two special issues: managing cultural diversity and managing "virtual" teams, where the team members are geographically separated.

The second chapter in this section, "The Challenge of Teaching Research Skills to Information Systems and Technology Students," was written by Beverley G. Hope of both Victoria University of Wellington (New Zealand) and City University of Hong Kong (China), and Mariam Fergusson, of PricewaterhouseCoopers (Australia). The chapter covers issues of teaching and learning graduate research skills. The authors identify the core research skills needed and present three pragmatic models for teaching them.

The last chapter of this section, "Data Modeling: A Vehicle for Teaching Creative Problem Solving and Critical Appraisal Skills," was written by Claire Atkins of the Nelson Marlborough Institute of Technology (New Zealand). Communication and problem-solving skills are fundamental for success as an IT professional. But can these skills be taught? Claire provides a method to teach these skills through the teaching of data modeling.

Section 2: Teaching Techniques and Pedagogy

The second section deals with teaching techniques and pedagogy or, more appropriately, with applying the results of countless research programs on how best to teach to our courses.

The chapter, "Towards Establishing the Best Ways to Teach and Learn about IT," by Chris Cope, Lorraine Staehr, and Pat Horan, all of Latrobe University, Bendigo, Australia, reports on an ongoing project to improve the ways they teach IT in an undergraduate degree. Using a relational perspective on learning, the authors developed a framework of factors to encourage students to adopt a deep approach to learning about IT. The authors define deep learning and describe the design, implementation, evaluation, and refinement of learning contexts and learning activities based on the framework they developed. Results are encouraging and show a significant positive effect when compared with a previous study by other researchers involving a different teaching and learning context.

In their chapter, "Computer-Supported Learning of Information Systems: Matching Pedagogy with Technology," authors Raquel Benbunan-Fich and Leigh Stelzer of Seton Hall University (USA) point out the need to apply new ways of teaching. This chapter proposes a three-dimensional framework to describe how best to teach, based on pedagogical assumptions of the given course, the time dimension of the communication between learners and teachers, and the geographical location of learners and teachers. The authors review the implications of the framework for IT education.

John Bentley and Geoff Sandy of Victoria University (Australia) and Glenn Lowry, United Arab Emirates University (UAE), present their chapter, "Problem-Based Learning (PBL) in Information Systems Analysis and Design," as the conclusion of this section of the book. This chapter looks at PBL as a possible way to provide a better match between information systems education delivery and the demands of the professional workplace. The chapter starts with a brief introduction to cognitive and learning principles, followed by a discussion of PBL and its potential to help to achieve a better fit between what students want and what employers demand. The chapter concludes with a concrete example of how to use PBL in a systems analysis and design course.

Section 3: Impact of the Web on IT Teaching

"Teaching or Technology: Who's Driving the Bandwagon?" by Geoffrey Mitchell of Victoria University of Wellington (New Zealand) and Beverley G. Hope of both Victoria University of Wellington and City University of Hong Kong (China) challenges our way of thinking about technology and education. Contrary to claims that the Web has revolutionized education, the authors argue cogently that many attempts at Web-based education simply reinforce current 'poor' teaching practices (perhaps disguised in new clothing). The authors argue that this occurs because of differing pedagogical assumptions and a limited understanding of how flexible learning differs from traditional approaches. They reason that flexible learning demands an increased focus on constructivism and the sociological aspects of teaching and learning. This chapter presents two frameworks that situate the authors' approach to flexible learning with respect to more traditional offerings and discusses the implications for educational technology design.

Karen S. Nantz and Terry D. Lundgren of Eastern Illinois University (USA) provide the chapter, "Delivering Course Material via the Web: An Introduction." The trend toward using the Web to deliver course material is ever increasing. Despite the many positive reasons for using the Web in a course, its use raises a number of new issues for faculty, including the time needed and salary concerns. The chapter presents a table of six levels of website use and discusses the major problems associated with creating and developing Web courses. The authors provide practical suggestions for dealing with these problems. The final section covers global enrollment, the electronic university, and the trend toward use of the Web for the delivery of course materials.

The chapter, "Bridging the Industry-University Gap: An Action Research Study of a Web-Enabled Course Partnership," by Ned Kock of Temple University (USA), and Camille Auspitz and Brad King, both of Day & Zimmermann, Inc. (USA), discusses a course partnership involving Day & Zimmermann, Inc. (DZI), a large engineering and professional services company, and Temple University. The course's main goal was to teach students business process redesign concepts and techniques. These concepts and techniques were used to redesign five business processes from DZI's information technology organization. DZI's CIO and a senior manager, who played the role of project manager, championed the course partnership. A website with bulletin boards, multimedia components, and static content was used to support the partnership. The chapter investigates the use of Web-based collaboration technologies in combination with communication behavior norms and face-to-face meetings, and its effect on the success of the partnership.

Section 4: Developing an IT Curriculum

The sections above have dealt with issues such as how best to teach a given course and topic. But these courses need to fit together to form a curriculum. This section deals with how to develop an IT curriculum that meets the ever-changing demands of industry.

Linda V. Knight and Susy S. Chan of DePaul University (USA) describe their efforts on "E-Commerce Curriculum Strategies and Implementation Tactics: An In-Depth Examination of DePaul University's Experience." Their master's e commerce curriculum was designed, developed, and implemented in just seven months. The program drew 350 students in its first year, and approximately 650 students with majors and concentrations in the e-commerce area in its second year. Their curriculum draws upon the principles of the IRMA / DAMA 2000, ISCC '99, and IS '97 model curricula. Strong technological expertise and infrastructure, solid industry relationships, and an entrepreneurial culture were critical success factors in developing and implementing the curriculum. The strategies that DePaul CTI employed and the lessons that it learned in the process of implementing its e-commerce curriculum are relevant to other universities seeking to move into the e-commerce arena.

Arthur Tatnall of Victoria University of Technology (Australia) and Bill Davey of RMIT University (Australia) point out how one can view "Information Systems Curriculum Development as an Ecological Process." Their chapter argues that if you want to understand *how* an IS curriculum is built, and how both the human and non-human interactions involved contribute to the final product, you need to use approaches that allow the complexity to be traced, and not diminished by categorizations or assumptions about intrinsic attributes of humans and non-humans. One way that this can be achieved is by using models and metaphors that relate to how people interact with each other, with the environment, and with non-human artifacts. This chapter uses the metaphor of ecology to achieve these goals.

Linda Knight and Susy Chan used various model curricula in designing their own. Anthony Scime of the State University of New York at Brockport (USA), in his chapter "Information Technology Model Curricula Analysis," puts these various model curricula into perspective. His chapter examines 10 model curricula on IT education and places them into grids that demonstrate how each curriculum model aims to educate students with various emphases on business, engineering, and mathematics.

The 2000 IRMA/DAMA Model Curriculum is our last chapter. It is the collective work of perhaps as many as 40 people, and I was pleased to serve as its editor. This document details an international information resources management curriculum for a four-year undergraduate-level program specifically designed to meet the needs of modern business. The curriculum provides a model for individual universities to tailor to their particular needs. That is, the IRMA/DAMA Curriculum Model is a generic framework for universities to customize in light of their specific situations. This curriculum model prepares students to understand the concepts of information resources management and technologies, methods, and management procedures to collect, analyze, and disseminate information throughout organizations in order to remain competitive in the global business world. These are all aspects of managing information. It outlines core course descriptions, rationales, and objectives, and includes suggested specific course topics and the percentage of emphasis.

This Curriculum Model addresses the needs of two distinct sets of learners:
1. students currently employed or seeking employment in the IRM field, and
2. all business students.

The IRM student needs specific in-depth understanding of IRM. All business students, if they are ever to manage effectively, require an understanding of how information management affects their job, the jobs of other managers, and their entire industry.

This book is current, relevant, and usable. Readers can start applying its lessons right away. It will prove invaluable to those revising their school's IT curriculum. Its concrete teaching ideas are perfect for those just starting out in teaching IT. Its coverage of curriculum issues from a global perspective is ideal for the experienced teacher in the throws of updating the curriculum. The book is also unique. It provides both a global perspective to IT education (so often lacking in other books) and advice based on solid research.

ACKNOWLEDGMENTS

This book could not have come to fruition without the dedication of many, many individuals. Let me start enumerating these dedicated souls by giving thanks to the members of its international board of reviewers. Each reviewer read and provided constructive feedback on three or more papers:

Louise Allsopp, Adelaide University (Australia)
Yvonne Antonucci, Widener University (USA)
Gene Aronin, National-Louis University (USA)
Clare Atkins, Nelson Marlborough Institute of Technology
    (New Zealand)
David Banks, University of South Australia (Australia)
John Bentley, Victoria University of Technology (Australia)

Khalid Benbachir, (Morocco)
Raquel Benbunan-Fich, Seton Hall University (USA)
Kathy Blashki/Sims, Monash University (Australia)
Dan Brandon, Christian Brothers University (USA)
Netiva Caftori, Northeastern Illinois University (USA)
Jeimy J. Cano, Newport University (Columbia)
Janet Carter, University of Kent at Canterbury (UK)
Frank Cervone, De Paul University (USA)
Dimitar Christozov, American University in Bulgaria (Bulgaria)
Tara Grey Coste, University of Southern Maine (USA)
Shimon Cohen, You-niversity.COM (Israel)
Chris Cope, La Trobe University (Australia)
Shirley Fedorovich, Embry-Riddle Aeronautical University (USA)
Meliha Handzic, University of New South Wales (Australia)
Alka R. Harriger, Purdue University (USA)
Igor Hawryszkiewycz, University of Technology, Sydney (Australia)
Jessie He Wong, (Canada)
Beverley G. Hope, Victoria University of Wellington (New Zealand)
Rodger Jamieson, University of New South Wales (Australia)
Gordana Jovanovic-Dolecek, Instituto Nacional de Astrofísica,
      Optica y Electrónica (Mexico)
Linda V. Knight, DePaul University (USA)
S. E. Kruck , James Madison University (USA)
Glenn Lowry, United Arab Emirates University (UAE)
Karen S. Nantz, Eastern Illinois University (USA)
Ian Newman, Loughborough University (UK)
Kathy Marold, Metropolitan State University of Denver (USA)
Mike Metcalfe, University of South Australia (Australia)
V.K. Murthy, Chinese University of Hong Kong (China)
M.V.Ramakrishna, Monash University (Australia)
Simos Retalis, University of Cyprus (Cyprus)
Celia Romm, Central Queensland University (Australia)
William T. Schiano, Bentley College (USA)
Anthony Scime, State University of New York at Brockport (USA)
Shi Nan Si, PhD IT Planning Manager, Singapore Pools Pte Ltd.
      (Singapore)
Lorraine Staehr, La Trobe University (Australia)
Tony Stiller, University of the Sunshine Coast (Australia)
Edward J Szewczak, Canisius College (USA)
Arthur Tatnall, Victoria University of Technology (Australia)
Winston Tellis, Fairfield University (USA)
Sal Valenti, Università degli Studi di Ancona (Italy)
Diane B. Walz, University of Texas at San Antonio (USA)

Edward Watson, Louisiana State University (USA)
Marilyn Wilkins, Eastern Illinois University (USA)
Connie Wells, Roosevelt University (USA)
Tongtao Zheng, University of Tasmania (Australia)

This book was able to benefit from experts around the globe by using a special paper review website. Thanks are deserved by Ketil Lund & Pål Halvorsen of Unik (Universitetsstudiene på Kjeller), who gave permission for us to use their ConfMan scripts for paper review. Ben Cohen, my nephew and Unix wizard extraordinaire, whipped my Web server into submission.

I thank Alex Koohang of National-Louis University who has been a continuing supporter of me personally and professionally. Let me again acknowledge the inspiration I have derived from the works of three individuals, whom I thank: Mehdi Khosrow-Pour, Nagib Callaos, and Olu Omolayole.

Special thanks go to the publishing team at Idea Group Publishing. Michele Rossi served as developmental editor. Her detailed lists of what to do, when, kept this project on track. Carrie Stull, Assistant Marketing Manager, had me complete a comprehensive questionnaire that provided the basis of the marketing campaign for this book.

This book has been graciously supported by Betty Boyd, my wife and partner. Betty provided constant support, particularly in the collation and blending of materials into a cohesive, informative format.

Eli Cohen
Editor

Section I

Examples on How to Teach Specific Topics

Chapter I

Teaching Teamwork in Information Systems

Connie E. Wells
Roosevelt University, USA

ABSTRACT

Teamwork is very important in information systems development. Therefore, most courses in systems analysis and design and many programming courses require students to work on group projects. However, a project group is not the same thing as a *team*. Furthermore, for a group to become a team, there are several important characteristics that must be developed. These characteristics do not always develop automatically. This chapter discusses the requirements for effectively forming, building, managing, and evaluating teams in information systems courses. Students should be taught these concepts in addition to the regular course content. This chapter also addresses two special issues that deal with team development and team management: managing cultural diversity and managing "virtual" teams, where the team members are geographically separated.

INTRODUCTION

Teamwork has been the norm for the development of many information systems (IS) projects since the early days of computing. Teamwork is also known to be an effective teaching and learning technique. Therefore, many IS classes in systems analysis and design, and also in programming, use student teams for class projects. However, just because students are required to do

Copyright © 2002, Idea Group Publishing.

their projects with other class members, it doesn't necessarily follow that they learn about effective teamwork. In fact, from previous experiences with classroom teamwork, many students dread taking another class that will require teamwork assignments. Why? It is because their experiences were filled with all sorts of problems and frustrations that they do not want to repeat. It is important, then, for many reasons, that IS instructors know about managing effective teamwork. And it is important that they *teach* teamwork in their courses, so their students will be prepared for effective teamwork in the "real world."

In 1999 an informal survey of the six top-selling college textbooks on systems analysis and design revealed that the word "team" or "teamwork" only appeared in the index of two of the books. These six books had a combined total of approximately four pages devoted to a discussion of teamwork or developing and managing effective teams. Informally, instructors will admit that one-fourth to one-half of their student project groups develop serious problems, and want to "fire" (or, "divorce") one or more of the group members. Obviously, those students have not learned how to work well in teams. In a "field quasi-experiment," Van Slyke, Trimmer and Kittner (1999) found that "... investing valuable class time on teamwork training can provide significant benefits. Student team members who receive teamwork training feel that their teams are more successful as a result" (p. 43).

The purpose of this chapter is to summarize what is known about effective teamwork, both in the "real world" and in the classroom. This chapter will also describe several specific techniques, activities, tools, and readings, which may be used in the classroom to *teach* teamwork rather than just assigning group projects to students.

BACKGROUND

Teamwork

What is a team? "The distinction between a work group and a team is an important one.... A work group becomes a team when shared goals have been established and effective methods to accomplish those goals are in place" (Wheelan, 1999, p. 3). A team, therefore, is more than just a group of individuals. For the purposes of this chapter, the following definition of a team will be used:

> "A team is a small number of people with complementary skills who
> are committed to a common purpose, performance goals, and ap-
> proach for which they hold themselves mutually accountable"

(Katzenbach and Smith, 1993, p. 45).

Working together as a team is more effective than working as individuals. "Teams outperform individuals acting alone or in larger organizational groupings, especially when performance requires multiple skills, judgments, and experiences (Katzenbach and Smith, 1993, p. 9). It is for this reason that thousands of books and articles on teamwork have been published over the past few years.

But teamwork in information systems is not new. In 1971 Gerald Weinberg wrote *The Psychology of Computer Programming*, in which he devoted a chapter to "The Programming Team." In the first paragraph of the chapter, he notes that, "The interaction of two programmers looking over a program that either one of them could have worked out is entirely different from the interaction of two programmers working on separate parts of a whole, which is too great for either one to produce" (p. 68). In 1998, he published a "Silver Anniversary Edition" of the book, which contained the original work with updated comments on each chapter. In these recent comments he says, "Over the years, I've observed that the requirement to develop capability cannot be met adequately by a single person. We learn much faster and much better with the active cooperation of others" (p. 5.i).

It has been traditional in most systems analysis and design classes, as well as many programming classes to assign students to group projects, in order to give students the advantages and experiences of teamwork. However, many instructors do not explicitly teach the principles of teamwork. Nor do they consciously encourage the development of the characteristics that are identified with successful teams and successful teamwork.

Successful Teams

What makes teams successful? Katzenbach and Smith's (1993) definition, quoted previously, included the following characteristics for a group to be called a team:

- Small size
- Complementary skills
- Commitment to a common purpose
- Have performance goals and approach
- Hold themselves mutually accountable (p. 45)

Other authors propose similar lists of characteristics that are required for successful teams. For example, Skopec and Smith (1997, pp. 12-15) identify six characteristics of what they call "Blue Chip Teams." These include:

- Identity
- Mission

- Great expectations
- Commonly accepted procedures
- Feedback
- Team-building

Hoffer et al. (1996) list the following as critical issues in teamwork:
- Selection of team members
- Having a common purpose (goal) and a commitment to the goal
- Mutual trust among competent team members
- Interdependence among team members
- Good communication among members
- Sense of empowerment and proper support from management
- Team-building skills (from training)
- Socializing and celebration

If students are to learn how to work effectively in teams, they need to understand these characteristics and the instructor should encourage the development of these characteristics in our student project groups. "Again, students need to be taught the interpersonal communication and team building skills that will help to ensure smoothly functioning groups. And we have a responsibility—if we expect them to work together—to provide that training" (Breslow, 1998).

Stages of Teamwork

Just as IS instructors identify multiple stages in a systems development life cycle (SDLC), teams are also thought of as developing in stages. In order to develop successful student teams, the instructor must communicate what the stages are, and encourage the successful development of each stage. Tuckman (1963) described a model of team development in four stages, naming them *forming, storming, norming,* and *performing*. His model is often cited in literature as well as in business. In Tuckman's model the *forming* stage is the initial orientation and establishment of relationships between members. The *storming* stage is "characterized by conflict and polarization around interpersonal issues." When the group moves beyond that conflict, new roles are developed and "ingroup feeling and cohesiveness" are developed; this is the *norming stage*. Finally, the group reaches the *performing* stage, when they develop roles that are more "flexible and functional," allowing them to achieve higher levels of performance (Tuckman, 1965, p. 396)

This chapter, however, has a slightly different model, because we are considering what stages a team will go through from the perspectives of the instructor and the students. The stages to be used in this chapter are: 1) Team

Figure 1: The Team Development Model (Source: Author)

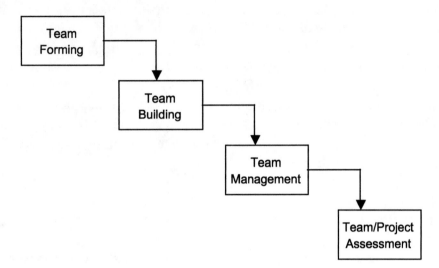

Forming, 2) Team Building, 3) Team Management, and 4) Team and Project
Assessment. Figure 1 models the four-stage model for a successful classroom
team project.

Stage 1: Forming Teams

In the first stage, the groups are formed. In business, managers will assign
employees to teams. Sometimes a team is formed just for one particular
project; other times teams are formed and will continue to work together on
many projects. In the classroom, student teams should be formed as early as
possible, to allow some time for the team-building stage, before the real work
of the projects begin. Successful student team formation requires that the teams
be of an appropriate size and that the selection of team members is done
according to consciously defined criteria.

Team Size

Katzenbach and Smith (1993) note that "virtually all the teams we have
meet, read, heard about, or been members of have ranged between two and
twenty-five people" (p. 45). In the classroom, student teams are usually three
to five members in size. If the group consists of only two members, there is the
risk of one member having to drop out, and the remaining student is left to do
the entire project alone. When the group is larger than five students, it becomes
significantly more difficult to coordinate the work and to ensure that all
members can arrange for a time to meet together. In addition, it is important
that all members can understand all parts of the project, and contribute equally

to the team. After all, the main purpose of the classroom group project is to have students learn the subject matter of the course, whether it is programming or systems analysis and design. Therefore, it is important that each member of the team be involved with, and understand, all parts of the project. A group that has more than five members makes this nearly impossible. Breslow (1998) suggests that, although four or five members are usually recommended, the number should be adapted to the amount of work to be done and to problems of logistics for the students. Howard (1999) suggests three to five members per classroom projects, and that "an odd number is recommended for greater success in working through conflicts."

Team Composition

There are several methods that are used to select the members of teams. Many instructors allow students to do self-selection. There are several arguments as to why this is not the best alternative. First of all, this is certainly not the way teams are formed in "the real world." Self-selected teams tend to be very homogeneous in composition. In one study, Baugher et al. (2000) found that self-formed teams will have less diversity of gender, race, and cultural background than groups that are randomly assigned. Although some would argue that homogeneity in teams will more likely result in less team conflict and better team performance, diversity in teams presents better learning experiences (Robbins, 2001). It may be argued that more conscious learning about teamwork can take place when students are not allowed to choose to work with the friends that they have worked with many times before.

Furthermore, as Katzenbach and Smith (1993) note, team members should have "complementary skills." These complementary skills for IS students will, naturally, include the students' backgrounds in terms of previous courses as well as their work experience in IS. Therefore, it is important that an instructor collect such information from the students. It is good to begin the course with a questionnaire that asks about experience with basic business software, programming languages, and the amount and type of relevant work experience that each student has. Students, then, may be placed into teams based on their knowledge and experience in the course subject matter. This would correspond with the concepts of chief programmer teams, as proposed by Mills (1970) and also described by Brooks (1995), where they describe programming teams using the analogy of surgical teams, where a highly skilled and experienced surgeon is supported by team members with other skills, and/or lesser skills. It also compares with systems development teams in businesses today, where a senior systems analyst is assigned to each team, to guide the less experienced team members through the project.

Personality characteristics are also important factors to consider when forming teams. Most educators are familiar with the *"Myers-Briggs Type Indicator®* (MBTI®) instrument, [which] is the most widely used personality inventory in history." (For more information on this, see the following website: http://www.meyers-briggs.com/products/mbti/index.asp.) This personality test may be used to help gain an insight into the differences between potential team members. Several other personality tests are presented and described in Robbins and Hunsaker (1996). These may be used to assess the students' personalities, which could be used as one of the criteria on which team composition is based. See Classroom Activity 1 for a suggestion on how to administer and use a personality assessment in the classroom. Whether or not students' personalities are part of the basis, or the whole basis, when forming teams, it is helpful for the instructor and the students to understand the variety of personality types and how this might affect intra-team communication and relationships.

Classroom Activity 1: Personality Styles

Have students take the "SAQ 4, Interpersonal Style Questionnaire," which may be found in Robbins and Hunsaker (1996, p. 18-30).* The questionnaire has 18 pairs of statements. Students are asked to rate themselves as to which of the statements in each pair is most descriptive of themselves. It only takes about 15 minutes for students to complete this questionnaire and then score their own results. These results will classify a student into one of four personality styles: *The Relater Style, The Socializer Style, The Thinker Style,* or *The Director Style.*

After students have rated themselves, have the students move so they are sitting in four groups, one for each of the personality styles.

Each group should discuss the four personality styles and develop a list of why they would want to be placed on a team that has a member from each of the other personality styles. For example, the *Relaters* should list why they would want to have a *Director* on their team, what the advantages would be in having a *Thinker* on the team, etc.

This activity will foster discussions that will encourage students to understand and to value the diversity of personalities and skills in team members.

*The SAQ is originally from Tony Alessandra and Michael J. O'Connor, *Behavioral Profiles: Self-Assessment* (San Diego: Pfeiffer & Company, 1994).

Other team member selection criteria

Criteria other than the students' IS skills and personalities may also need to be considered for practical reasons. These criteria include the students' work/study schedules, and their geographical proximity or distance from other students. These factors can present logistical problems for a student team. If these are significant criteria for the students at your school, you could create a blank daily schedule form. Each student should mark on the schedule the times when he/she cannot meet with a team. With different symbols, they could identify when would be the best times to meet with their team and other times that, perhaps, would not be best, but would be possible. It is important to recognize that many of our "non-traditional students" are working full-time while they are attending classes. Thus, their schedules may be at least as important as other criteria.

Another option in forming the teams would be to collect information in all of these areas: skills, personality, and schedules. Then, with all factors in a spreadsheet, the instructor could work through, making compromises where they seem most appropriate.

Stage 2. Building Teams

"Team building refers to activities aimed at enabling a group to become a cohesive working unit capable of functioning at the highest performance levels. Effective team building helps a team establish an appropriate organization and work culture, and accelerates the accumulation of experience in functioning as a team" (Weinberg, 1995, p. 38). Assigning students to a project group does not automatically, overnight, make them into a team. To have truly effective *teams* instead of typical classroom work groups, there should be some conscious effort at team building. Harper and Rifkind (1995) list several guidelines for team development. These are listed in Table 1: Guidelines for Team Development, and are discussed in the following paragraphs.

1. *Get to know one another.* In order for there to be good communication and mutual trust among team members, they must have some means of getting to know one another and to "break the ice." The most obvious thing that students need to do first is to learn each others' names, exchange addresses, telephone numbers, etc.
2. *Seek ways to "connect" team members.* The students should be encouraged to at least briefly introduce themselves to each other, perhaps sharing some of their previous educational and work experience, family background, and so on.

Some classroom exercises may be used for the ice-breaking as well as to illustrate points later in the course. See Classroom Activity 2: Breaking the Ice.

3. *Develop a team vision.* In this step the team members need to begin to communicate with each other about their vision, goals, and objectives relative to the course and the group project. To encourage the group to form a common purpose (goal) and a commitment to the goal, various authors, including Marble (1992, 1993), suggested that the group begin by developing a common identity. The first group assignment may be to decide on a name for the group, as if they were a consulting company. They should design a logo and the letterhead for their correspondence, and they should write a mission statement.

4. *Develop a group character.* With the mission statement, the team is establishing standards of quality for their team work. The group should also plan for their meetings. When will they meet? Where will they meet? Each student should sign a commitment that he/she will hold to the quality standards set forth in their mission statement and allocate meeting times, as planned.

Students should be assigned readings, or given handouts, that describe the various roles that need to be established for an IS team to be effective. See Table 2: Student Roles in Teams for a suggested list of roles and activities that may be required on an IS team. The members might assign individuals to those roles, or adapt the roles to fit the number of students in the group and the talents and personalities of the members. They may also write job descriptions for the roles and résumés for the team members. Whether or not these roles are formalized and maintained, it is

Table 1: Guidelines for Team Development*

1.	Get to know one another.
2.	Seek ways to "connect" team members.
3.	Develop a team vision.
4.	Develop a group character.
5.	Create a context that is safe for team participation.
6.	Discuss each of the phases of team development as they occur.
7.	Develop task processes
8.	Establish mechanisms for team self-assessment and improvement.
9.	Find ways to celebrate the team and its accomplishments.

*From Harper & Rifkind, 1995, pp. 66-68.

Classroom Activity 2: Breaking the Ice

In *People and Project Management*, Rob Thomsett (1980) describes "The Tinkertoy, Game." This is an exercise that "simulates a typical user, analyst, and programmer situation." A target model (swing set, for example) is constructed ahead of time and placed in a nearby room where it is not visible to the class. Each team is given a package of Tinkertoy pieces* from which they will be asked to reproduce the target model. There is a race to see which team can replicate the model in the shortest time.

One team member is the *user*, who will be allowed to see the target model and will then have to describe the model to the *systems analyst* so the systems analyst can relay that information to the *programmers*. The programmers then have to build the model.

As described by Wells (1999), this activity may be used to illustrate several concepts:

1.   *Communication is difficult.* The *user* may have a very clear picture of what he/she wants, but it is very difficult to communicate exactly what is desired.

2.   *Drawing system models helps with communication.* At first, the communication is only verbal. When they are allowed to draw pictures, the task is made much easier.

3.   *Decomposition is important in complex projects.* The most successful teams will use a top-down approach in describing the model and its parts.

4.   *Analysis before development.* In the most successful teams the *analyst* takes a long time with the *user* before running to communicate with the *programmers*. It is important to fully understand what the *user* wants before beginning construction.

This exercise can help the team members work together, get to know one another, and have fun together, before beginning the real "work" of the class project.

*Note: Tinkertoy sets are relatively difficult to find nowadays. K'NEX®, are the modern-day equivalent, and may be found in most toy stores. For more information on this activity, see Wells (1999).

helpful for the students to know what responsibilities are necessary to have a successful team project.

5. *Create a context that is safe for team participation.* Harper and Rifkind (1995) propose a code of ethics for teamwork. Their code of ethics is briefly listed in Table 3. It would be good for the students to read this code of ethics and adopt it, or adapt it, for their own group. The activity of discussing the code of ethics and of addressing the issues and conflicts that can arise in group projects will be a positive step in the team-building process.

6. *Discuss each of the phases of team development as they occur.* If the students are guided into reading and discussing articles on team development, and if they understand that Tuckman's (1965) *storming* is a normal

Table 2: Student Roles in Teams

Presider/Meeting Leader
  • Keeps group's meetings on task
  • Develops and distributes an agenda for each meeting
  • Monitors the group's progress

File Manager/Project Master
  • Keeps assignment/project files secure; makes backup copies
  • Makes sure all members have current copies of assignments
  • Coordinates and integrates the project components
  • Maintains frequent contact with the other group members for
  project updates

Meeting Coordinator
  • Knows the schedules of all team members
  • Decides the dates and times of team meetings
  • Notifies members of scheduled meetings
  • Has the authority to call a meeting if no conflicts exist with any member's
      official schedule

Intermediary
  • Acts as the primary contact between the group and the course instructor
  • Meets periodically with the instructor to report on group progress
  • Should keep aware of how the team is progressing on the project and
  whether there are any major conflicts between group members

stage for teams, it will help them understand what is happening. Team members should view diversity and some conflict as an asset. The personality inventory results and the activities suggested earlier can help to begin the process of valuing diversity. Other comments on diversity and cultural differences will be discussed later in this chapter. Having the students read articles that discuss the values of different viewpoints, the advantages of diversity, will help. When conflict arises, it will be comforting for the students to know that this is a normal part of the team development process. Managing team conflict will be discussed in a later section. For this guideline, though, students should be encouraged to keep a journal. Although it may be tempting for them to just record project data in the journal, a journal can serve a much better purpose if it is used to record team process observations. Students should record dates and decisions, meeting attendance, and other facts in the journal. However a journal is an especially good place to record the difficulties encountered in making decisions, feelings about the team and its processes, and other items of introspection about teamwork.

7. *Develop task processes.* After assigning roles for the team members, there should be agreements about how some of the tasks will be done. For example, who will pay for photocopies? What about the team project notebook? How will they communicate with each other, outside of meeting times? These may seem like minor decisions, but they help to increase communication and develop more of an identification with the team.

8. *Establish mechanisms for team self-assessment and improvement.* In this stage, the instructor can guide students. It is often difficult for students

Table 3: Code of Ethics for Teamwork*

1. Team members should accept responsibility for their own actions.
2. Team members should act so that the potential for future teamwork is maximized.
3. Team members should maximize individual freedom of choice.
4. Team members should act so that the respect of each participant for self and for others is maximized.
5. Team members should act to improve communication.
6. Team members should view diversity as an asset.

* From Harper and Rifkind, 1995, p. 66.

to honestly and tactfully address their weaknesses and how to overcome problems. In the stage "Managing Teams" (below) are some suggestions that the instructor may use to give the students a start on discussing performance issues.

9. *Find ways to celebrate the team and its accomplishments*. Although many college students will not need encouragement to socialize and celebrate, other teams become so intense and have so much stress from their work, classes, and family obligations, that it is very difficult for them to celebrate. But this also will help to increase the team spirit, their unity and identity. Let the students know that it is good to celebrate accomplishments, to have social times as well as work times.

Stage 3:   Managing Teams

From the lists of characteristics of successful teams, it is obvious that team members must share a common goal or mission, have a defined set of procedures, and feel a sense of empowerment and proper support from management. In this regard, the instructor should act as a good business manager. That is, he/she should give clear assignments, describe the project work that needs to be done, when each part of the project is due, and be available to give direction when it is needed. However, to encourage the sense of empowerment, the instructor should take care not to "micro-manage" student teams.

There are several types of short questionnaires that may be used to help team members gain insight into their communication patterns, responsibilities, and inter-personal skills within the team. For example, after the teams have been working on their project for a while and have completed some of the project requirements, a questionnaire dealing with team leadership could be developed. The questionnaire could be in the form of a table. The first column would list the activities that are needed by successful teams. (See the bulleted items in Table 2: Student Roles in Teams.) Each student fills in the table, with the name of each team member as a column heading. The last column can be assigned to "No One." For each activity, the student should assign a value of "1" in the column of the member who mostly does that function. A "2" could be entered for another team member who also contributes to that activity. The last column will exist in case *no one* has been doing that activity.

Another questionnaire, with a similar tabular design, could list the specific project assignment activities and deliverables, broken down into detail. For example, if the teams have completed the project planning phase, the list might include 1) conceptual planning of the activities, 2) constructing the PERT chart, 3) typing the report, etc. As in the students' questionnaire about roles, numerical ratings could be entered into columns to show which student(s) have worked on each part of the project task. A "1" would be assigned to a team member that worked alone on that task. A "2" would be assigned to each of two or more team members who shared responsibility for that task. When this assessment is done several times over the life of the project, the work levels of the team members will become more balanced.

A third questionnaire might ask each student to rate, on a scale of 1 to 5, the team climate, including concepts such as openness, support for individuals, addressing conflicts, etc. There are many short questionnaires dealing with such topics in Robbins and Hunsaker's book (1996).

All of these questionnaires (on communications, tasks, etc.) can help the instructor diagnose when problems are developing and help teams design methods of solving those problems. As was stated earlier, an instructor should not micro-manage the student teams, but as a manager, the IS instructor should meet with teams one or more times during the course to check on how the teamwork is progressing. If the instructor just meets with a team and asks, "How is it going? Are you all getting along okay? Any problems?" the answers will nearly always be just general reassurances that things are okay. By using specific questionnaires, which are completed by the individuals and submitted to the instructor before he/she meets with the teams, there will be specific topics that may arise and need to be discussed. In fact, just the activity of completing the questionnaires will cause the individuals to assess their roles and activities. They will become aware of the amount of work some members are neglecting, while others must carry most of the load.

Managing Conflict

Conflict is a part of teamwork. This is the *storming* stage of Tuckman (1965). The question is not whether there will be conflict, but how to address the conflict and have it not escalate into large battles. A positive view of conflict is given by Katzenbach and Smith (1993), in which they state:

Conflict, like trust and interdependence, is also a necessary part of becoming a real team. Seldom do we see a group of individuals forge their unique experiences, perspectives, values, and expectations into a *common* purpose, a set of performance goals, and approach without encountering significant conflict. And the most challenging risks associated with conflict relate to making it constructive for the team instead of enduring it (p. 110).

To address the ways to manage conflict, it may be helpful for the instructor to read some articles on conflict resolution. Have students also read them, so their awareness will be heightened and they will be given positive ways of addressing conflict. Suggested sources include Harper and Rifkind (1995), Wheelan (1999), Skopec and Smith (1997), Robbins and Hunsaker (1996), and Katzenbach and Smith (1993). Skopec and Smith (1997) list several views of group conflict and offer several suggestions for conflict management (pp. 87-98). Lewis (1993) also has a large part of his book devoted to understanding team dynamics and managing team conflict. Assigning an article or two on conflict resolution may also enable students to admit that they have problems and give them a safe climate in which they can discuss positive solutions. There are also excellent videos on conflict resolution and assertiveness training that are available from libraries or commercial sources.

Stage 4: Team and Project Assessment

Virtually all classes require that the instructor assign grades. This means that an assessment must be made for the teams/ team projects. Several issues need to be discussed with regard to assessment of the teams and/or projects. Young and Henquinet (2000) discuss team project evaluation in terms of a two-by-four matrix. The first question that must be addressed is the *purpose* of the assessment. Is the purpose only to evaluate the results of the teamwork—the project deliverables? Or, is the purpose to evaluate how well the students worked together as a team? Young and Henquinet (2000) describe this as an issue of *product* versus *process*. The instructor my choose to evaluate both product and process, depending on the goals of the class. If class goals include both learning how to do systems analysis and design or about writing good software (i.e., *product*) and learning about learning group dynamics, communicating well with other team members, attending all meetings, etc. (the *process*), then these goals should be stated in advance and should be given appropriate weights in the assessment process.

Another question that needs to be asked is whether all students on the same team should be given the same grade. Some authorities argue that if the team is to be committed to a common goal and to work cooperatively toward that goal, it is important for them to know that they will be evaluated as a team. As Skopec and Smith (1997) say,

"One of the seminal articles in the literature of compensation management is entitled 'On the Folly of Expecting A While Rewarding B,' by former Academy of Management president Steve Kerr. His point is at once simple and profound: companies cannot hope to direct work behaviors in one direction while the compensation system directs them in another direction" (p. 123).

If students expect to be evaluated as individuals, rather than as a team, there will be some students who will intentionally take on more work than is fair.

Some have even been known to call team meetings at times they know other members cannot meet. This will ensure that the work and credit that might have been shared with other team members will be done only by themselves, so they can honestly claim to have done most of the work on the project. Such attitudes do not encourage good teamwork.

On the other hand, it must be acknowledged that there are, at times, freeloaders, or slackers, on student teams. If they are to be evaluated as a team, some students will enjoy the free ride, and let the other team members work for the grade that they will all be given. Is it fair to have one student pass the course because their (other) team members were intelligent, hard workers, while he/she did nothing? Kagan (1995) argues strongly that group grades are not fair, and can cause a number of problems for the instructor.

What is the solution? The arguments for a common team grade and the arguments for individual grades all seem reasonable. A compromise may be made. For example, the grade may have about one-fourth of the grade allocated to the individual's participation. Have each student assign a portion of the "individual effort points" to each of the team members. For example, if there are four team members and 100 points is a perfect score for the team, then 25 points for each team member would indicate that they all cooperated well and the work was evenly divided among the four. If one member has not contributed his/her share, then that member's points may be given to other team members, arguing that overall the project score would have been higher, had that member done as well as the other members. Be sure that each team member adds comments to the evaluation to support his/her assignment of the points. Experience shows that there is usually a high level of agreement among the team members on this score. This type of scoring rewards teams that have worked well together, divided up the work, and given each other the support that they needed. It also leaves the remaining 75 percent of the team's score to be a group score, to encourage the high performance of the team as a whole.

Colbeck et al. (2000) discuss other alternatives, such as dividing the project into parts, some of which should be done by individuals, and other parts that should be done by the group. Then the course will have evaluations based on both individual and team performance.

The final question, the other dimension of the Young and Henquinet (2000) matrix, is that of *who* should do the evaluation. They identify four possibilities. In most traditional courses, it is the instructor, alone, who must evaluate both the team and the resulting project. However, team members may also be asked to evaluate themselves and the other members of their team. A third option is for those who are classmates, but outside the team, to evaluate the team (product, usually not process, in this case). A fourth possibility is to have the team projects evaluated by people outside the classroom. The teams may be asked to give presentations and documentation to other faculty, or to local IS managers, so they may also be involved in the evaluations. If the team has been working on a project for some department or organization outside the class-room, representatives of those units should also be asked to evaluate the project.

Regardless of who all are involved in the assessment process, the team members and all evaluators should be given the evaluation criteria and the forms ahead of time. This will make sure that everyone is aware of the grading criteria. Of course, in order for the team members to have established the proper common goals, they should have been aware of the entire evaluation process from the beginning of the project.

CURRENT AND FUTURE TRENDS

Two trends that already have considerable influences on our teamwork and our classrooms are cultural diversity and the use of technology to enable classes and teamwork to take place over long distances. Because they offer

Table 4: Seven Cultural Dilemmas*

1. **Universalism vs. Particularism**. Refers to whether the culture focuses on rules that apply to everyone, or if particular circumstances are more important than the rules
2. **Individualism vs. Collectivism**. Is quality of life defined by the conditions of individuals or by society as a whole?
3. **Neutral vs. Affective Relationships**. Do we display emotions to others?
4. **Specific vs. Diffuse Cultures**. Refers to how long it takes to be allowed into a person's private areas, or how large the private areas are for individuals.
5. **Achievement vs. Ascription**. How does a person get respect and status? Does it come from accomplishments, or does it come from age, class, gender, etc.?
6. **Internal vs. External Orientation**. Does the person believe that he/ she can control everything, or do nature, events, and other people control the individual?
7. **Past-Oriented vs. Future-Oriented**. Relates to the time dimension. This has to do with respecting the past, looking toward the future, keeping schedules, etc.

* Adapted from Berger, 1996, pp. 18-28.

other dimensions to developing and managing teams, they are discussed separately in the following sections.

Managing Cultural Diversity

One issue that has been gaining more attention over the past few years is cultural diversity in society, in universities, in businesses, and, therefore, in teams. Culture may be defined as a type of collective programming of the mind. This programming begins as soon as a baby is told what to do and what not to do. Harper and Rifkind (1995, p. 29) use Dodd's (1991) definition of culture: "The total accumulation of an identifiable group's beliefs, norms, activities, institutions, and communication patterns."

Berger (1996) describes seven dimensions of culture. These dimensions deal with three areas: "our relationships with other people, our relationships with nature, and our relationship with time." These seven dimensions he presents as dilemmas that affect our management style, human relations programs, and relating to others in a group. Table 4 lists and briefly describes the seven cultural dilemmas. Berger (1996) describes each of these dilemmas in more detail and then discusses management applications and ways of reconciling the differences.

The most recent U.S. Census report confirmed what most communities already knew, that diversity has significantly increased in the past few decades. In recent years there has been such a shortage of programmers and others with special skills in information technology, that immigration quotas for the United States have been adjusted to bring in more of these skills from other countries. Most of the universities with significant numbers of students in IS and/or computer science will also have an increasingly diverse range of cultures, both in the students and in faculty members. And, even if our students don't physically leave their community, the Internet and its applications have made computing to be an international meeting ground. The applications our students will be developing will be much more oriented to regions beyond where they physically live and work. Therefore, there is an increased demand on teams to understand and manage cultural differences.

Cross-Cultural Work Groups (1997), edited by Granrose and Oskamp, is devoted to discussing the various aspects of cultural diversity and the influence of cultural diversity on work groups. This diversity affects interpersonal relations, job satisfaction, and work productivity. Recognizing cultural differences is only one step toward avoiding or resolving cultural conflict.

The teamwork code of ethics cited earlier, exercises in valuing diversity among team members, and the already suggested readings in conflict management will help to set a climate of understanding among individuals, regardless of their cultural differences.

Managing Teams Across Distances (Virtual Teams)

The second recent issue that is increasing in importance is that of managing distance among team members. Many more schools are offering "distance learning" as part of their curriculum. Businesses are often multinational, with a need to have teams in more than one country. How we can build and manage project teams that may have members spread out over large distances? Although we may also call these "virtual teams," that label may be misleading. The teams are *not* virtual, in the sense that they are *real*. For them to be real teams, they must go through similar development stages and provide the same advantages as do our face-to-face, classroom-based student teams. Lisa Kimball (1997) cites several things that "need to happen in order for organizations to make effective use of virtual teams" (p. 1). These things are listed in Table 5.

Many of our traditional classes are learning to use virtual classroom technology to support and enhance their traditional teams. Several forms of media may be used. They include video conferencing, audio conferencing, electronic mail, Internet-based "chat rooms," and asynchronous Web conferences. These may be used to support student project teams, whether or not the

Table 5: Requirements for Effective Use of Virtual Teams*

• Processes for team management and development have to be designed, defined, piloted, tested, and refined
• Team managers have to be trained in new team management strategies
• Team members have to be trained in new ways of working
• The culture of the organization has to be reshaped to support new structures and processes
• Organizational structures have to be modified to reflect new team dynamics
• Rewards systems have to be updated to reflect new team structures
• New information technology (IT) systems have to be built to support teams
• New management, measurement, and control systems have to be designed
*From Kimball, 1997.

team is part of a class that is offered totally as distance learning.

There are several companies that provide Internet-based classroom support. Using these online systems, classes may be totally Web-based, or may be traditional classes that use the Internet only for enhancements. One example is Blackboard®. An instructor can register and set up a course on Blackboard.com, and then post class documents, announcements, and other materials using this Web-based service. Within a course, the instructor can also create groups of students. Students can either schedule "chat" sessions, post asynchronous communications and documents, and/or use Blackboard®, to send each other e-mail with attachments. Students can use their "drop-box" to turn in assignments to the instructor, without the other students or other teams being able to access and see the documents.

Aside from services such as Blackboard®, universities can also set up remote access for faculty and students, so there can be individual and/or team access to CASE tools on a university server.

But is technology enough? Obviously, from Table 5, we see that many changes must take place for virtual teams to become effective. The instructor must see that those changes are made so that the technology can enable or support the forming, building, and managing of effective team projects.

Trust among members of a team is very important. Coutu (1998) cites studies that indicate that virtual teams do not develop trust in the same patterns that face-to-face teams do. With virtual teams, the trust factor must be established immediately, rather than developing slowly. The interactions between virtual team members must begin immediately, and they must begin with social messages, introducing themselves to the others. Since there may not be body language nor voice tones and inflections to give members clues to one another, initial messages must be more personally informative.

In addition, the researchers stated that following the initial personal messages, there is a need to "set clear roles for each team member" (Coutu,

Table 6: Strategies for Effective Virtual Teams*

• Make whole visible to everyone
• Provide "line of sight"
• Catalyze rich conversations
• Amplify energy
• Create tracks and footprints in physical space
* From Kimball, 2000.

1998, p. 20ff). This allows each team member to know who to contact for a certain activity, and what will be expected of them by the other team members.

In a speech given at a Team Strategies Conference sponsored by Federated Press, Toronto, Canada, Kimball (1997) describes several strategies that "make a significant difference in team effectiveness. These strategies are listed in Table 6.

The ideas of *wholeness* and *visibility* are critical. The team must somehow *feel* like a team, rather than separate individuals. She suggests that the team brainstorm on ways to create this "wholeness." The brainstorming idea itself will make members aware of the importance of thinking as a team, rather than individuals. Suggestions that have been made include posting a team photograph by each member's computer, like on a mouse pad or a calendar, creating and distributing a map of where group members are, or creating a graphic that has meaning to the group.

The idea of *lines of sight* is to encourage group communications that can, somewhat, replace face-to-face group meetings. When a group has face-to-face meetings, there will usually be verbal reports, written reports, and minutes of the meeting. Again, Kimball suggests having the group brainstorm about how they can create substitutes for these lines of sight using modern technology.

Similarly, Kimball suggests brainstorming on ways that the group can *enrich conversations* with graphics, models, pictures, etc. Attention must also be given to ways the group can maintain and *amplify energy*, so the group spirit and enthusiasm don't die down. She also notes that there must be a formal way to *create tracks and footprints in physical space*. How will records be kept? Who will keep the records? In what form? Using the team name, team logo, and other items created in early assignments will be constant reminders that the work is to be teamwork, rather than work done by a group of individuals.

CONCLUSION

Learning how to build, manage, and participate in teams is a very important success factor in careers in developing information systems. If instructors in these disciplines are to prepare students for their careers, there must be a conscious effort to have students both informed and experienced in working in teams. As cited in the introduction, Van Slyke, Trimmer, and Kittner (1999) found that students do, indeed, benefit from training in teamwork. They conclude saying,

Many IS programs have responded [to employers] by integrating team projects into their curriculum. However, it may not be enough to simply throw students together and expect them to function effectively as a team. Specific training in teamwork skills may be a necessary component for successfully enabling students to be effective team members (p. 44).

Finally, a few other points should be made. First of all, remember that an important part of team building and team management is socializing. Socializing helps build communication, trust, and commitment. At various times during a project, it is important to take time out to celebrate and feel good about the teamwork and the job. Students should not be discouraged from having a good time while they are working on team projects.

And, although this chapter emphasizes the skills and techniques that encourage successful teams and team projects, the instructor must remain aware of the main subject matter of his/her course. The students need to learn the theory, concepts, and methodologies of systems analysis and design and/ or programming. Most of the suggestions in this chapter take very little classroom time. Only a total of about two to three class hours, at most, need be spent directly on teaching teamwork. But those hours can provide both the student and the instructor with a lot of payback, in terms of student learning and successful team projects.

REFERENCES

Alessandra, T., and O'Connor, M. J. (1994). *Behavioral Profiles: Self-Assessment*. San Diego: Pfeiffer & Company.

Baugher, D., Varanelli, Jr, A., and Weisbord, E. (2000). Gender and culture diversity occurring in self-formed work groups. *Journal of Managerial Issues,* 12(4), 391-402.

Berger, M. (1996). *Cross-Cultural Team Building: Guidelines for More Effective Communication and Negotiation*. London: The McGraw-Hill Companies.

Breslow, L. (1998). Teaching teamwork skills. *Teach Talk Articles in the Faculty Newsletter* X(4). [Online]. Available: http://web.mit.edu/tll/published/teamwork1.htm.

Breslow, L. (1998). Teaching teamwork skills, part 2. *Teach Talk Articles in the Faculty Newsletter* X(5). [Online]. Available: http://web.mit.edu/tll/published/teamwork2.htm.

Brooks, F. P., Jr. (1995). *The Mythical Man-Month: Essays on Software Engineering* (20th anniversary edition). Reading, MA: Addison-Wesley.

Constantine, L. L. (1993). Work organization: Paradigms for project management and organization. *Communications of the ACM*, 36(10), 35 - 43.

Coutu, D. L. (1998). Trust in virtual teams. *Harvard Business Review*, 76(3), 20-21.

Dodd, C. H. (1991). *Dynamics of Intercultural Communication* (3rd edition). Dubuque, IA: William C. Brown Publishers.

Gardner, B. S., and Korth, S. J. (1999). A framework for learning to work in teams. *Journal of Education for Business,* 74(1), 28-33.

Granrose, C. S., and Oskamp, S. (Eds.) (1997). *Cross-Cultural Work Groups*. Thousand Oaks, CA: Sage.

Harper, L.F. , and Rifkind, L. J. (1995). *Cultural Collision: Quality Teamwork in the Diverse Workplace.* Dubuque, IA: Kendall/Hunt Publishing Company.

Hoffer, J.A. et al. (1999). *Modern Systems Analysis and Design* (2nd edition). Reading, MA: Addison-Wesley.

Howard, S. A. (1998). Guiding collaborative teamwork in the classroom. *Effective Teaching*, 3(1). [Online] Available: http://cte.uncwil.edu/et/articles.htm

Kagan, S. (1995). Group grades miss the mark. *Educational Leadership,* 52(8), 68-71.

Katzenbach, J. R., and Smith, D. K. (1993). *The Wisdom of Teams: Creating the High-Performance Organization*. New York: Harper Business.

Kimball, L. (1999). Managing virtual teams. (Text of speech given for Team Strategies Conference sponsored by Federated Press, Toronto, Canada, 1997.) [Online]. Available: http://www.tmn.com/~lisa/vteams-toronto.htm.

Kimball, L. (2000). *Ten Key Elements for Team Leaders to Manage.* [Online] Available: http://www.caucus.com/pw-tenkey.html.

Lankard, B. A. (1994). Cultural diversity and teamwork. *ERIC Digest No. 152*. ERIC Identifier: ED377311, 1994-00-00. [On-line] Available: http://www.ed.gov/databases/ERIC_Digests/ed377311.html.

Larson, C. E. and LaFasto, F. M. J. (1989). *Teamwork: What Must Go Right/What Can Go Wrong.* Newbury Park, CA: Sage Publications, Inc.

Lewis, J. P. (1993). *How to Build and Manage a Winning Project Team.* New York: American Management Association.

Lipnack, J., and Stamps, J. (1997). *Virtual Teams: Reaching Across Space, Time, and Organizations with Technology.* New York: John Wiley & Sons, Inc.

Marble, R. (1992). *Casebook for Systems Analysis and Design: F.S.S., Inc.* New York: Mitchell McGraw-Hill.

Marble, R. (1993). *Casebook for Systems Analysis and Design: J.P.S., Inc.* New York: Mitchell McGraw-Hill,

Mills, H. D. (1970). *Chief Programmer Teams: Techniques and Procedures*. IBM Internal Report, January.

Montebello, A. R. *Work Teams That Work: Skills for Managing Across the Organization*. Minneapolis, MN: Best Sellers Publishing, 1994.

Parker, G. M. (1990). *Team Players and Teamwork: The New Competitive Business Strategy*. San Francisco: Jossey-Bass Publishers.

Robbins, S. P. and Hunsaker, P. L. (1996). *Training in Inter-personal Skills: Tips for Managing People at Work*. Upper Saddle River, NJ: Prentice Hall.

Robbins, T. L. and Fredendall, L. D. (2001). Correlates of team success in higher education. *The Journal of Social Psychology*, 141(1), 135.

Skopec, E. and Smith, D. M. (1997). *The Practical Executive and Team Building*. Lincolnwood, IL: NTC Business Books.

Thomsett, R. (1980). *People & Project Management*. Englewood Cliffs, NJ: Yourdon Press.

Tuckman, B. W. (1965). Developmental sequence in small groups. *Psychological Bulletin* 63(6), 384-399.

Van Slyke, C., Trimmer, K., and Kittner, M. (1999). Teaching Teamwork in Information Systems Courses. *Journal of Information Systems Education*, 10 (3&4), 36-45.

Weinberg, G.M. (1971). *The Psychology of Computer Programming*. New York: Dorset House Publishing.

Weinberg, G.M. (1998). *The Psychology of Computer Programming* (Silver anniversary edition). New York: Dorset House Publishing.

Wells, C. E. (1999). A team building and communications exercise for the systems analysis and design class. In Reither, B. J. (Ed.), *Proceedings of the Decision Sciences Institute Southwest Region,* March 10-13, pp. 129-131.

Wheelan, S. A. (1999). *Creating Effective Teams: A Guide for Members and Leaders*. Thousand Oaks, CA: Sage Publications.

Yamane, D. (1996). Collaboration and its discontents: Steps toward overcoming barriers to successful group projects. *Teaching Sociology,* 24(October), 378-383.

Young, C. B., and Henquinet, J. A. (2000). A conceptual framework for designing group projects. *Journal of Education for Business,* 76(1), 56-68.

Chapter II

The Challenge of Teaching Research Skills to Information Systems and Technology Students

Beverley G. Hope
Victoria University of Wellington, New Zealand
and City University of Hong Kong

Mariam Fergusson
PricewaterhouseCoopers, Australia

As the information systems discipline grows, so do the number of programs offering graduate research degrees. These include one-year post-graduate (honors) programs, masters by research, and doctoral degrees. Graduate students entering their first research degree are faced with a quantum leap in expectations and required skills. The burden is significant: they need to find a referent discipline, select a research method and paradigm, defend the research relevance, and fulfill the requirements of adding to a body of knowledge. The purpose of this chapter is to inform discussion on the issue of teaching and learning graduate research skills. We identify the core research skills needed and present three pragmatic models for teaching them. This provides a basis for a shared knowledge and discourse based on lessons learned.

Copyright © 2002, Idea Group Publishing.

INTRODUCTION

Student experiences of conducting information systems and technology (IST) research in Australia generally begin in graduate programs. At the undergraduate level, IST education focuses on fundamental concepts, applications, and skills for practice. This vocational education acts as a terminating point for many students, but an increasing number are continuing or returning to graduate education in order to broaden their options in the labor market. Many of these graduate programs include a research component and, to meet the needs of employment-focused students, this research needs to be both rigorous and relevant.

The question is, how do we train people to become good researchers? What are the required knowledge, skills, and abilities of a good researcher? And how do we foster these in our students? If these are cognitive skills, *can* they be taught and learned? We believe the answer to the last question is a firm "yes." We learned to research; our students also can learn. But how best do we achieve this?

The objective of this chapter is to examine issues of how to foster research skills in novice IST research students. We propose a minimal set of skills and understandings (thought processes and concepts) as part of the researcher's repertoire, and outline some teaching strategies that can foster their development. The important assumptions underlying the chapter are:

- research is situated, that is, what constitutes knowledge and good research are defined by the audience or discipline; and,
- research skills are not innate, they are learned (and, consequently, can be taught).

THE IMPERATIVE FOR TEACHING RESEARCH

Research requires a spectrum of cognitive abilities, from the simple ability to establish facts to the more complex ability to judge and evaluate. In higher education we might reasonably expect students to be led through this spectrum to the point where they are able to criticize, to analyze, and to reach a deep understanding of knowledge. Typically, however, students are graduating from their first degrees in IST with limited research training and under-developed critical thinking skills. One reason for this may be that undergraduate IST education is seen as a professional qualification rather than the first step in an academic career. The need for practitioner skills and the ever-increasing knowledge base in IST leaves little time for research skilling. At the undergraduate level, we teach *what* is known, not *how* it is known. Furthermore, an increasing number of students are entering graduate programs with strong professional experience in lieu of academic prerequisites.

Figure 1: IST research: A juggling act

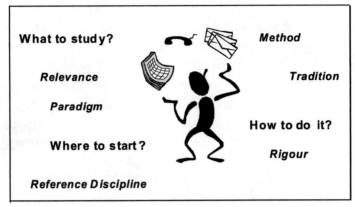

So, IST students are entering graduate research degree programs without understanding the research process and without possessing research skills.

Graduate research degree programs vary in content and structure, but the output invariably includes a written document, whether a research project, published paper, thesis, or dissertation. The document is the culmination of many months or years of study and serves not only to communicate findings but also to evidence the student's ability to conduct credible research. To be credible in the IST field, the research must be both relevant and rigorous.

Relevance and rigor are paradigmatic issues because our world views guide both the selection of phenomena to study and the acceptability of methodologies (Wand & Weber, 1986; Weber, 1997). As such, paradigms set the scene for the selection of a *good* topic (relevance) and provide a benchmark or standards for evaluation (an aspect of rigor). In contrast to established disciplines, IST is at best multi-paradigmatic and at worst non-paradigmatic. It remains to a large degree a complex "fusion of behavioral, technical, and managerial issues" (Keen, 1980, p.10), that are difficult to untangle. The discipline borrows theories, models, and methods from the physical and social sciences, as well as from professional business disciplines. As early as 1980, Keen noted the plethora of reference disciplines in practice, including computer science, experimental social psychology, cognitive psychology, and political science (Keen, 1980, p.10). There are no fewer today. The 1993 *Keyword Classification Scheme for IST Research Literature*, for example, contains 16 reference disciplines (Barki, Rivard, & Talbot, 1993). It is not surprising, then, that Galliers' (1991, p.328) review of the literature identifies 14 research approaches used in IST. With no single accepted paradigm, multiple reference disciplines, and a variety of research approaches, IST researchers must make some important decisions up front. They must select an "appropriate" discipline from which they can set standards, draw theories, select methodologies, and find criteria for evaluating research (Keen, 1991). The novice IST researcher is on a very steep learning curve.

Research students in more mature disciplines do not normally share the ambiguities facing novice IST researchers. For example, science and engineering students have a well-defined research paradigm that is both normative and scientific. At masters and PhD levels, these students are likely to become part of a research team with an established area of study. The *apprentice within a team* approach provides for a rapid start, an effective mentoring system, and the benefits of collegiality. Typically, IST students entering research programs do not have these advantages. The lack of a team support structure is particularly marked under the British or Oxbridge PhD tradition followed in most of Australasia. The structure imposed by course-led programs such as those in North America mitigates this to some extent. We use the term 'course' here and throughout this chapter to denote a unit of study, sometimes referred to as a paper, unit, or subject.

The early stage in the research life of a student has been identified as being critical (Fergusson, 1997). In an empirical study students reported a big leap in expectations and skill requirements upon entry to an honors (first graduate research) program, but found the step up to PhD study less problematic (Mullins & Kiley, 1998). The difference between these steps is knowledge and experience of research. A rapid start is particularly important for honors students who have a short time to complete their program of study, but is also important for masters and PhD students seeking a timely completion date. To ease students into research, we argue that they need to begin their graduate research degree programs with research training.

TEACHING AND
LEARNING RESEARCH SKILLS

Research skills, like any other skill, can be learned. Just as students once learned to analyze an information system and draw a DFD, so they can learn to analyze literature and write academic articles. Even creativity can be learned (Gerity, 1997). But how are research skills best acquired? While there is a plethora of books to assist students in completing theses (see, for example, Creswell, 1994; Leedy, 1997), these usually address the research process rather than the component skills. It has even been argued that textbooks can be a barrier to creative thinking as they "instill judicial thinking" (Evans, 1991, p.47). The term judicial thinking implies a correct way of critiquing, judging, or solving problems. Thus books, while providing a useful framework for the process of producing the research output, do not replace the valuable contribution of the research instructor, mentor, or class.

Teaching research skills represents a shift from focusing on the research topic to focusing on learning outcomes. Student learning outcomes "encompass a wide range of student attributes and abilities, both cognitive and affective" (Baddeley, 1979, p.4). In describing doctoral research training, the Australian Vice-Chancellors' Committee noted that it:

> Provides training and education with the objective of producing graduates with the capacity to conduct research independently at a high level of originality and quality. The student ought to be capable by the end of their candidature of conceiving, designing, and carrying to completion a research program without supervision (Australian Vice-Chancellors' Committee [1990] cited in Brown [1998]).

A subsequent review modified and softened this definition of traditional research training:

> Research training, through good supervision, should provide the student with an ethical framework and generic workplace competencies as well as research-related skills.... Some universities have embraced this broader concept of research training through the establishment of graduate schools which appear to be more focused on the multiple dimensions of research training and the different career paths that research training provides (Review Committee [1997, p.3] cited in Brown [1998]).

The differing approaches to research education can be arrayed on a continuum ranging from the lone scholar approach to the student cohort approach. At the lone scholar extreme, novice researchers work largely on their own to prove themselves capable of self-skilling in the research process and of providing a unique contribution to knowledge. At the other extreme a cohort of students are led through the research process in a series of classes or workshops with ongoing peer and mentor support.

The lone scholar approach is the traditional form of research training in Australasia as it is in the United Kingdom. Here, students work with supervisors in a one-to-one relationship. In theory, this provides students with a way of internalizing the work habits and expertise of a seasoned researcher by association; in reality, the student will rarely observe their supervisor in action. As Wood-Harper, Miles, and Booth (1993, p.455) observe, "The culture of individualism within UK academia is such that post-graduate students are expected to make their own way in the research world."

There are many possible learning outcomes at this end of the training continuum. Learning can be serendipitous, and dependent upon both the problems that the student encounters and the skill and attitude of the supervisor. If the supervisor is a subject but not a research specialist, acquisition of research skills can be minimal. The success of the relationship can be affected by the chemistry of the personality mix and the workload or availability of the supervisor. At one end there are students who are left to "muddle along" and teach themselves research skills. They learn from books or from emulating a method from published research. Learning is most often just-in-time, exposing the student to the risk of learning too late that an essential step was omitted earlier in the process, perhaps invalidating results. While they may become adept in the chosen method, they are unlikely to be knowledgeable in a variety of methods or to even know of their existence. These researchers acquire a narrow view of the research process and are likely to become method rather than problem driven. The implication of method-driven research is that knowledge is circumscribed by the method by which it is acquired. The lone scholar approach emphasizes depth at the expense of breadth and individualism at the expense of collegiality and debate (Baddeley, 1979; Davies, 1990).

The self-taught student is the poor cousin in the research arena, because, as Baddeley (1979, p. 130) observes, "Most subsequent occupations are likely to involve interacting with other people, writing in such a way as to interest and influence them, and meeting reasonably short deadlines. The PhD student is given neither training nor experience in any of these."

The problems of the lone scholar may be overcome by the collegiality, support, and structure at the cohort end of the training continuum.

The cohort approach is more common in the United States, but is fast being taken up by other regions, including Australasia. Students learn about a variety of research methods through a combination of attending seminars, analyzing published research, completing practical assignments, and writing papers. Course content may be dependent upon the prevalent paradigm of the particular IST department or the bias of the instructor. Some address only positivist paradigms and are heavily quantitative, but increasingly the interpretivist paradigm and qualitative methods are being taught. The coursework mode fails to realize its potential when the curriculum favors a particular paradigm or methodology, or fails to include experiential learning.

Constructivist theory of knowledge suggests that actual knowledge is not transmitted from one person (the teacher) to another (the student); rather it is constructed in the mind of the student (Cronin, 1997; Jonassen et al., 1995). This is achieved by situating learning in real-world contexts and grounding concepts in experiential learning. In research education, this involves practicing research skills on real or realistic research problems with all their complexity, to provide context to the process, thereby enhancing learning. Students are then more likely to be able to articulate and clarify concepts, and to develop critical thinking abilities (Metz & Tobin, 1997).

While all modes may lead to success, we propose that more benefits are obtained at the cohort end of the continuum. Coursework provides a focused and structured learning environment, ensuring better coverage of a broad range of research paradigms and methods. In addition it is more economical both of staff and student time. By providing the right kinds of experiences, we can enhance our student's ability to search the literature, think critically, and write concisely in an academic style. They can learn how to challenge assumptions, critique methodologies, and improve article organization. A particular strength of the research classroom is that students are able to learn from each other's experiences as well as their own. This enriches their view of the approaches to research, and allows them to work together on essential skills, but independently on their own research projects. They also learn through debate to articulate and defend their argument, an important skill. In the long term, the structured teaching of research methods and skills must benefit the IST discipline as graduating students go on to conduct research and disseminate their findings.

To construct a research training program, we need to be able to identify the skills and techniques used by experienced researchers. These provide a framework around which research programs can be organized.

CRITICAL RESEARCH SKILLS

For the purpose of identifying critical skills it is convenient to conceptualize research as a linear series of steps grouped according to the common milestone outputs. In reality, the research process is likely to involve several iterations and include feedback loops (Barrow & Thomson, 1997). From the literature (see for example, Kumar, 1996) and from our experience, we posit the following process and outputs:

Phase I - Topic Analysis
1. Choose a topic area
2. Survey the literature (and practice)
3. Identify a relevant research problem/opportunity
4. Identify the audience for the intended research output
5. Identify the research approach suited to the problem and the audience

Phase II - Research Proposal
1. Thoroughly search the literature and synthesize the findings
2. Define the research questions, with hypotheses where required by the paradigm

3. Develop a research design and method, including a time plan for the research
4. Identify key assumptions and limitations

Phase III - Research Report
5. Collect and analyze "data"
6. Interpret results in light of existing knowledge and theory
7. Communicate the findings (thesis, paper, article)

These skills can be grounded in the student's own research. Outputs at the end of each phase are the practical applications of the lessons learned in the student's actual research: topic analysis, research proposal, and research report (thesis, conference paper, or journal article). The outputs provide the formal framework within which the students' understanding can be expressed and subsequently discussed as a cooperative group learning experience.

This process suggests a "minimum" set of critical skills that the novice researcher must acquire to complete good research. These are shown in Table 1 under the headings "Cognitive skills" and "Research techniques." Although the distinction between cognitive skills and research techniques is not clear cut, the separation is intended to provide a classification against which to target particular areas for development and to compare the offerings of existing courses.

Reading Table 1 across and then down provides a loose matching with the research process steps set out earlier. The six essential skills can be used to benchmark existing or proposed courses for novice researchers. The skills can be taught to a group while allowing sufficient diversity for individual development. In addition, the list of skills can be used as a reference for criterion-based assessment.

Because relevant research frequently involves study of IST within organizations, we can add to the required skill set, consultancy skills. According to Metzger (1993), these include: diagnostic ability, problem-solving skills, specialized

Table 1: Essential Research Skills

Cognitive skills	Research techniques
Selecting a topic and defining a research problem	Conducting a comprehensive search of the literature
Evaluating and synthesizing existing research	Designing the research and "data" collection
Selecting and justifying a research paradigm	Writing an academic paper

knowledge, communication skills, marketing and selling abilities, and business acumen.

RESEARCH PROGRAMS

Programs of study for research students tend to vary considerably. They vary both in terms of the relative proportions of coursework to research and the amount of prescribed research methods coursework. For example, at Victoria University, honors study begins with an intensive one-week program (Lowry, 1997); at the Australian Defence Force Academy, students take one research methods course; and at Victoria University of Wellington, honors students take four research-focused courses. The objective of each program is to focus students on defining a research question, developing a research plan, and conducting credible research. These three programs are summarized next.

Case 1: Honors at Victoria University (VU), Victoria, Australia

Lowry (1999) reports a semi-structured approach to research education at VU. The approach was developed and successfully implemented earlier while Lowry was at the University of Tasmania (see Lowry, 1997). Of the 22 students participating in the research training program at VU in 1998, 17 were PhD candidates, three were DBA candidates, and two were master's students. Of the 17 PhD candidates, 13 were serving members of the academic staff. The nature of the student body, particularly the high proportion of academic staff enrolled, favored an intensive, workshop approach to research training.

Program aims included development of an appreciation of the broad nature of the IS discipline, an awareness of the range of research methods, an ability to articulate research questions, and competence in "framing, conducting, arguing, and reporting research in a convincing yet conventional way" (Lowry, 1999). Additional objectives included the rapid completion of PhDs, particularly by serving academic staff, and increased publication output from the PhD research.

Training begins with an intensive, week-long orientation in which participants begin their thesis with identification of a preliminary research question and development of a working title. The initial workshop is followed with three workshops, at monthly intervals. These workshops, which are "social as well as academic events", are held in an executive seminar setting and are characterized by strong student input and interaction. In describing the earlier, Tasmanian, program, Lowry (1997) describes the workshop structure as:

1. In the first follow-up workshop, students report their progress and focus on the literature review and research questions. They are expected to have substantially completed their focused reading and to have written a draft literature review.

2. In the second workshop students report progress, focusing on refinement of the research question, hypotheses, scope, and assumptions. They are expected to have completed the thesis introduction (Chapter 1) and literature review (Chapter 2).

3. In the third workshop, students focus on research design and plans. They are expected to have fully revised previously presented chapters, and to have a good first draft of the methodology (Chapter 3).

In describing the subsequent VU program, Lowry reports that at the end of the first semester, students are expected to have developed a formal research proposal and substantially completed the first three to five chapters of their thesis. Outcomes of the training have included increased staff publications, higher PhD completions, and development of three research groups within the department-organizational, technology, and technology-induced change management,

Case 2: Honors at the Australian Defence Force Academy (ADFA), University College, UNSW Canberra, Australia

Students enrolled in the honors program at ADFA are full-time students and personnel in the Australian Defence Force. Most have limited or no prior research training.

The aims of research training are to facilitate completion rates, decrease student frustration, and improve the quality of research outputs. Students are required to take four courses and a thesis over one calendar year, with Research Methods prescribed in the first semester. A course is the equivalent to 36 contact hours. Each student is allocated a research supervisor.

The assessments for the Research Methods course are: topic analysis, literature review, research presentation to the school, and a research proposal. Assessment is criterion referenced and marked independently by the course lecturer/manager and the student's supervisor. The role of the supervisor is to provide topic-specific expertise and advice, including help with finding the relevant literature, and choosing a suitable research paradigm.

Students are encouraged to find a topic themselves. Students who ask their supervisors for a research question tend to be more organized with their work and able to focus early on the body of knowledge surrounding their topic. Their enthusiasm for the topic is typically high initially, and reduces gradually over the year. Students who elect to find their own topic are slower to "settle," though they tend to maintain momentum until the end of the year. There is no perceived difference between the final performance of students who choose their own topic and those who obtain one from a supervisor.

The identified cognitive skills are covered in the course material and practiced in assignments. Workshops and mini-tasks are used to assist the students with selecting a topic and defining a problem, as well as evaluating and synthesizing existing research. Research paradigms used in IST are discussed in lectures, and justification of the research paradigm is a required part of the research proposal. Training in research techniques tends to be limited to writing skills, and this mostly by feedback on submitted work. Students have an on-line resource center, which "guides" them through the research process as presented in the course. The greatest weakness of the students by the end of the semester-long research methods course is in providing alternate research designs.

Following the introduction of the research methods course, formative evaluations were requested of research supervisors. These consistently reported significant improvement in student performance, better understanding of the topic areas, and perceptions of greater levels of motivation among students. Feedback obtained from around 25 students over a three-year period was positive with respect to their own research. Students were all able to identify direct benefits of having completed the course. About 20% of students reported that "learning about all the 'other stuff' (research paradigms that they did not use) was useless." It would seem that grappling with the breadth of the IST discipline, on top of learning how to do research, might be a little too much to ask of an honors student in a single semester, though the vexatious question of how to expose them to the range of paradigms and methods remains.

Case 3: Honors at Victoria University of Wellington (VUW), Wellington, New Zealand

(Despite name similarity there is no connection with Case 1). The honors year in the School of Information Management at Victoria University of Wellington also serves as the first year of a two-year master's-by-research degree. An important objective of the IST honors program is to provide the foundation skills needed for students to progress to the master's thesis and eventually, we hope, to a doctoral degree.

The majority of students are full-time, recent graduates of the school's undergraduate program, but a few graduates from other universities in New Zealand and overseas join the program. In any given cohort, one or two students are likely to be part time, while also working downtown or in the home. Whatever the circumstance, students generally have little prior academic research experience.

The program of study consists of eight courses, of which four are prescribed. Elective courses are content specific and include options in areas such as e-commerce, the virtual organization, and IT in the new organisation. Two electives may be chosen from cognate disciplines, and common choices include marketing, management, and library and information science.

All four prescribed courses are devoted to research issues and practice. They are: Foundations of Information Systems Research, Research Methods in Information Systems, Current Issues in Information Systems Research, and Research Project in Information Systems.

Foundations of Information Systems Research is a first-semester course, which introduces the IST discipline and includes ontological foundations of IST, an overview of the main streams of IST research, and the skills and techniques required to write a literature review. A skills workshop on searching the university online catalog and online databases is provided. Assessments include bi-weekly article critiques, bi-weekly search and writing exercises, and an academic literature review. Topics for the literature reviews are selected from a list of topics to be offered by staff as research projects in the following semester. In this way, students get an early start on their second semester research project.

Research Methods in Information Systems is also a first-semester course, which provides a critical examination of methodologies used in IST research. Both qualitative (case research, grounded theory, ethnography, action research) and quantitative methods (survey, experimental study) are included. Skill workshops include introductions to NU*DIST and SPSS. Assessment includes weekly assignments describing research methods and reviewing related articles, a research journal describing the student's learning process, an end-term test, and a research proposal. The research proposal is the major course output and brings together the skills and knowledge of the course. It also provides some depth in an otherwise broad coverage of research methods. Students are encouraged to write their research proposal for their proposed project in the following semester, and to this end they incorporate into the proposal the literature review completed and edited in the *Foundations* course.

Current Issues in Information Systems Research in the second semester builds on the *Foundations* course. It involves a critical examination of recent literature in the domain of strategic, managerial, and organizational aspects of IST research. Relevance of research is a strong theme in this course. The issue of relevance is also addressed through the incorporation of a practitioners' forum. Assessment includes bi-weekly article reviews, two mini-projects or literature reviews, and an end-term test.

The Research Project in Information Systems provides an opportunity for students to synthesize the learning in all other courses and prepares them for entry into Part II of the master's program, the thesis. Students work with a selected supervisor on a project offered by that staff member. In our first year of offering the program, we allowed students to select and define their own research question, but we found that students were taking too long, often half the semester, to define their topic. Since honors is a one-year, eight-course program, this was clearly unacceptable and the quality of some projects reflected this. Since moving to the 'select from a list' approach, students have completed the assignment in a more timely manner, projects have substantially improved, and students are less stressed. Under the revised scheme, students can refine the research question but they are helped by the provision of a general research topic. A conference-style presentation to staff and students is required at the end of the course.

The VUW honors program is thus an intensive introduction to research incorporating both cognitive skills and research techniques. Students work long hours and burnout is a potential problem. The program does not meet every entrant's needs. Those who enroll expecting to receive consultancy training are disappointed, but those who enroll for the intended research degrees are well served. We have been impressed by the quantum leap in understanding and practice evidenced by our students following on to theses from the honors program. Many have subsequently published their research projects.

These three research training programs are not intended to be prescriptive. Rather, they are examples of approaches designed to meet the differing needs of differing cohorts at different institutions. We hope that they will foster debate and serve as starting points for others to develop (and report) their own research training initiatives. The reviewed programs had similar aims and all assumed a cognitive apprenticeship model within the constructivist educational paradigm. All programs were considered successful by participating teachers, research supervisors, and research students.

CONCLUSION

As a discipline without an accepted single paradigm, IST presents novice researchers with many challenges. The applied nature of the discipline imposes the dual requirements of rigor and relevance, which can be difficult to meet even for seasoned researchers. Consequently, IST students entering graduate research degree programs benefit from structured research training experiences. Decomposing the research process into a set of critical skills can assist in developing research training programs and criterion-referenced assessments.

To demonstrate the applicability of the skill-set, we described three quite different programs, each of which were considered to meet the needs of their constituents. VU offers an intensive, week-long introduction, followed by three, day-long workshops; ADFA, offers a single, semester-long course which combines theory with practice; and VUW offers a series of courses. Courses include both cognitive skills and research techniques, though techniques are more frequently practiced in the longer training program at VUW.

Further research is needed to explore the different modes by which students acquire research skills and to assess the relative effectiveness and efficiency of each. Such research could help us to identify which modes are best suited to which learning aims and objectives. From our experience and from our review of the literature, we recommend that any research training be given a strong contextual grounding. The skills taught must be relevant to both future academics and future practitioners.

REFERENCES

Australian Vice-Chancellors' Committee (1990). *Code of Practice for Maintaining and Monitoring Academic Quality and Standards in Higher Degrees.* Canberra: AVCC.

Baddeley, A. (1979). Is the British PhD system obsolete? *Bulletin of the British Psychological Society, (32),* 129-131.

Barki, H., Rivard, S., & Talbot, J. (1993). A keyword classification scheme for IS research literature: An update. *MIS Quarterly, 17*(2), 209-226.

Barrow, P. D. M., & Thomson, H. E. (1997). Choosing research methods in information systems. In *Proceedings of the 2nd UKAIS Conference* (pp. 239-251). Southampton, United Kingdom.

Brown, D. (1998). The PhD as professional qualification: Is research education enough? In Kiley, M. and Mullins, G. (Eds.), *Quality in Postgraduate Research: Managing the New Agenda.* Adelaide: University of Adelaide. Available at: http://www.usyd.edu.au/su/supra/quality.pdf (last accessed March 12, 2001).

Creswell, J. (1994). *Research Design: Qualitative and Quantitative Approaches.* Newbury Park: Sage Publications.

Cronin, P. (1997). *Learning and Assessment of Instruction.* Web document. http://ww.cogsci.edu.ac.uk/~paulus/Work/Vranded/litconsa.htm (last accessed March 8, 2001).

Davies, G. (1990). New routes to the PhD. *The Psychologist, 3*(6), 253-255.

Evans, J. R. (1991). *Creative Thinking in the Decision and Management Sciences*. Cincinnati, OH: South-Western Publishing Co.

Fergusson, M. (1997). The Ferrett: A Web tool for postgraduate supervision. In Sutton, D. (Ed.), *Proceedings of the 8th Australasian Conference on Information Systems,* (pp. 529-539). Adelaide, Australia.

Galliers, R. D. (1991). Choosing appropriate information systems research approaches: A revised taxonomy. In Nissen, H. E., Klein, H. K., & Hirschheim, R. (Eds.), *Information Systems Research: Contemporary Approaches and Emergent Traditions* (pp. 327-345). North-Holland: Eslevier Science Publishers.

Gerity, B. P. (1997). *Teaching Creativity: A Cognitive Apprenticeship Approach to the Instruction of Previsualization in Art.* Web document. http://cc.usu.edu/~slqxp/TCResear.html (last accessed March 15, 2001).

Jonassen, D., Davidson, M., Collins, M., Campbell, J., & Haag, B. (1995). Constructivism and computer-mediated communication in distance education. *The American Journal of Distance Education, 9*(2), 7-26.

Keen, P. (1980). MIS research: Reference disciplines and a cumulative tradition. In Mclean, E. (Ed.), *Proceedings of the First International Conference on Information Systems* (pp. 9-18). Philadelphia, USA.

Keen, P. (1991). Relevance and rigor in information systems research. In Nissen, H. E., Klein, H., & Hirschheim, R. (Eds.), *Information Systems Research: Contemporary Approaches and Emergent Traditions* (pp. 27-49). Amsterdam: North Holland.

Kumar, R. (1996). *Research Methodology: A Step-by-Step Guide for Beginners*. Melbourne: Addison Wesley Longman.

Leedy, P. (1997). *Practical Research: Planning and Design*, 6th ed. New Jersey: Prentice Hall.

Lowry, G. R. (1997). Postgraduate research training for information systems: Improving standards and reducing uncertainty. In Sutton, D. (Ed.), *Proceedings of the 8th Australasian Conference on Information Systems* (pp. 191-202). Adelaide, Australia.

Lowry, G. R. (1999). Building a research culture at the Victoria University greenfields site. *Australian Council of Professors and Heads of Information Systems (ACPHIS) Curriculum and Research Workshop*, Macquarie University, NSW, September 30-October 1, 1999 (invited paper).

Metz, P., & Tobin, K. (1997). Cooperative learning: An alternative to teaching at a medieval university. *Australian Science Teachers' Journal, 43*(1), 3-28.

Metzger, R.O. (1993). *Developing a Consulting Practice*. Newbury Park: Sage.

Mullins, G., & Kiley, K. (1998). Quality in postgraduate research: The changing agenda. In Kiley, M., and Mullins, G. (Eds.), *Quality in Postgraduate Research: Managing the New Agenda* (pp. 1-13). Adelaide: University of Adelaide.

Review Committee. (1997). *Learning for Life–review of Higher Education Financing and Policy: A Policy Discussion Paper*. Canberra: AGPS.

Wand, Y., & Weber, R., (1986). On paradigms in the IS discipline: The problem of the problem. In *Proceedings of the 1986 Decision Sciences Institute Conference* (pp. 566-568). Honolulu, Hawaii.

Weber, R. (1997). *Ontological Foundations of Information Systems*. Melbourne: Coopers & Lybrand.

Wood-Harper, A.T., Miles, R.K., & Booth, P.A. (1993). Designing research education in Information Systems. In Khosrowpour, M., & Loch, K. D. (Eds.), *Global Information Technology Education: Issues and Trends* (pp. 453-487). London: Idea Group Publishing.

Chapter III

Data Modeling: A Vehicle for Teaching Creative Problem Solving and Critical Appraisal Skills

Clare Atkins
Nelson Marlborough Institute of Technology, New Zealand

INTRODUCTION

Despite extensive changes in technology and methodology, anecdotal and empirical evidence (e.g., Davis et al., 1997) consistently suggests that communication and problem-solving skills are fundamental to the success of an IT professional. As two of the most valued skills in an IT graduate, they should be essential components of an effective education program, regardless of changes in student population or delivery mechanisms. While most educators would concur with this view, significantly more emphasis is generally placed on teaching the tools and techniques that students will require in their future careers, and a corresponding amount of energy is expended in attempting to identify what those tools and techniques might be. In contrast, successful problem solving is often seen either as an inherent capability that some students already possess or as a skill that some will magically acquire during the course of their studies.

Data modeling as an activity, by which we mean the gathering and analysis of users' information needs and their representation in an implementable design, is largely one of communication and problem solving and, consequently,

Copyright © 2002, Idea Group Publishing.

provides an excellent opportunity for explicitly teaching these skills. Data modeling is generally considered to be one of the more difficult skills to teach (e.g., Hitchman, 1995; Pletch, 1989), particularly if the student has no previous understanding of physical data structures (de Carteret & Vidgen, 1995). The essential constructs, such as entities, attributes or objects, may be elegant in their powerful simplicity, but their combination into a useful design is a complex process of categorization in which there is "considerable room for choice and creativity in selecting the most useful classification" (Simsion, 1994 p.82). Data modeling requires not only the ability to communicate about and to solve a problem, but also to create possible solutions and then choose between them. Herein lies the difficulty. It is not enough to learn what the different constructs are, or even to study simple textbook examples of how to put them together. The student must really understand the problem, be able to create and recognize a number of possible ways in which the problem can be solved, and then exercise considerable critical skills in choosing between them.

This chapter examines these issues and describes various ways in which final-year undergraduate students, taking a specialist module in data modeling, have been encouraged to develop, and have confidence in, their creative and critical ability to solve problems in a disciplined and systematic way.

BACKGROUND

In many respects the entire systems development lifecycle (SDLC) can be considered as a complex problem-solving activity. The need to communicate effectively with all levels of users, with technical personnel and with management has long been accepted as an integral part of the process. Texts on systems analysis and design generally include at least one chapter on the topic, including basic skills of interviewing, report writing and presentations. These texts will often use the language of problem solving to describe the activities and context of the SDLC, for example referring to the system environment as the 'problem domain,' the need to 'identify the problem' and 'seeking solutions.'

In their exemplary text, Satzinger et al. (2000) speak specifically of the "Analyst as a Business Problem Solver," commenting that analysts must have both "a fundamental curiosity to explore how things are done and the determination to make them work better" (p.4). However, while they diagrammatically describe the SDLC in problem-solving terms, the book is primarily focused on detailed and clear explanations of tools, techniques, methods and methodologies, with little indication of the thinking skills that are required to synthesize information, to extract the problems and to create alternative solutions. Yet it is precisely these skills that are required to transform a novice into a competent analyst or designer.

Clearly, learning the tools of the trade is an important and necessary requirement of an IT student's education but this alone is not sufficient. It can be argued that it is the remaining work that a student engages in, particularly assignments, that provides these necessary skills but is this really the case? In the 'real world' there are no model answers, no completely right or wrong solutions. However, there is a limit to the amount of subjectivity we can allow our students in a modeling, analysis or even programming assignment if we are to have any chance of providing fair and timely feedback. This conundrum is highlighted by Simsion and Shanks' (1993) experiment where each of 51 modelers, with varying experience, provided a different data model solution to a given scenario, and only three were more than superficially similar. Consequently, it seems far more effective, and certainly easier, to provide students with a limited and tightly focused case study that provides only limited opportunity for interpretation and creativity but which, by focusing primarily on the correct use of tool or technique, can be graded with some confidence. However, we need to question whether this approach is providing students with the best possible skill set.

The debate about which programming languages or development methodologies should be taught in depth are only likely to become more complex and more protracted as the IT profession continues to diversify. It is no longer possible to second-guess whether a student will be best served by learning Java or C++, Rational or ORACLE; any choice is likely to be outdated within a short time. Different institutions or instructors will make different decisions on these matters, based on a number of factors, but underlying any of these choices must be a commitment to developing the students' ability to ultimately identify the problems, find the solutions and choose for themselves. Communication studies and, less often, problem-solving skills may be taught as discrete modules in some programs but all modules need to acknowledge these skills and integrate them with the more technically specific teaching topics. In this way, students not only increase their opportunity to practice these skills but also to see how relevant they are to all subject areas.

As a specific teaching area, data modeling has some inherent difficulties of its own and many of them are related to the wider problems of teaching any design technique. As Batra and Zanakis (1994) point out, there is not a "precise set of rules and heuristics to develop an E-R diagram"(p.228) because "most design methods are informal in the way in which the task of identifying candidate objects is achieved" (Eaglestone and Ridley, 1998). This can result in the phenomenon that Pletch (1989) observed where modeling students often display the same characteristics of high school math students who "sit and look at the words in the text for some time waiting for the required equation to be miraculously revealed to them" (p.74). In order to make any move towards a useful design, a novice modeler has to come to a clear understanding of the 'problem' situation, has to test out a number of possibilities, often by trial and error, and then use some ill-defined set of criteria to choose between them. Even then the choice may not be straightforward. As Simsion and Witt (2001) emphasize, there is definitely "more than one workable answer in most practical situations" (p6).

Given the difficulty of recognizing the 'best' solution for anything other than trivial examples, together with the obvious problem of determining what criteria a student has used for classification, and consequently how a student has arrived at a solution, it is difficult for an educator to provide any objective assessment or feedback. Once again the easier option of restricting the assessed scenario and focusing the grading on the 'correct' use of constructs and notation can seem very attractive and more impartial. However, once again, it is at the expense of encouraging the development of the very skills that the expert modeler needs. Indeed, it can be counter-productive, encouraging students to concentrate on discovering the (assessor's) expected solution rather than experimenting with different and possibly innovative solutions (Postman and Weingartner, 1969).

In order to address some of the issues, a specialized data modeling module was developed in 1994 for use with final-year undergraduate Information Systems and Computer Science students at Massey University in Aotearoa/ New Zealand. The students had received standard tuition in systems analysis and database design, and were very familiar with the constructs of entity-relationship (ER) and relational data modeling. However, they had had little opportunity to experiment with using these techniques in scenarios of their own choice. The module was designed specifically to redress this lack and to provide a wide variety of opportunities to try out a number of techniques that could potentially assist them to improve their thinking, critical appraisal and communication skills and, above all, to bolster their confidence in these abilities.

DATA MODELING TECHNIQUES–A TAUGHT MODULE

Module Overview

The module, named Data Modeling Techniques, was developed as a single-semester elective module consisting of approximately 160 hours of student learning of which a third was timetabled contact time with the teacher (i.e., lectures and tutorials). The module was described as "the study and application of techniques commonly used in the business world for formalizing information needs and data structures, together with an investigation into the importance of data modeling as a means of communication and understanding." Between 1995 and 2000, the average number of students taking the module was 40, which represented about 40% of the students graduating with a major in Information Systems.

Module Objectives

The objectives as laid out in the module documentation were:

- to appreciate the value and purpose of data modeling as a business technique,
- to appreciate several aspects of the problem-solving process,
- to thoroughly understand the techniques of entity-relationship/relational modeling as commonly used in business situations,
- to gain a holistic view of the data modeling process,
- to experiment with ways of applying the techniques, and
- to successfully apply all the above to a problem domain of interest to the student.

In addition, it was also explained to students that expertise in data modeling could not come from study alone and that none of them would be 'experts' at the conclusion of the module. Instead, it was emphasized that expertise comes largely from experience, from exposure to a wide range of modeling applications and situations, and from the ability to think creatively 'outside the square.' It was also explained that it was not possible to gain this kind of expertise without making mistakes and that a failed exercise was often a positive learning opportunity. Consequently, the module had a strong emphasis on rewarding demonstrated experiential learning rather than 'right' answers. In addition, the module explicitly explored the kinds of mistakes, which research suggests, are made by novice modelers and focused on strategies for avoiding them.

Module Structure

The module ran in four hours of timetabled contact time with the instructor per week. Of these, one hour was used for the delivery of lecture material, one hour for the presentation of student seminars and two hours for interactive tutorial activities. Students were also expected to spend eight hours each week either on work set as part of the summative assessments or in reading and class preparation. The schedule of assignments is shown at Table 1. For most of the

Table 1: Assessment Schedule

Assignment	Weight	Purpose
Problem Solving: You will be asked to solve up to five problems requiring the use of imagination, lateral thinking and advanced problem solving strategies.	15%	To encourage you to think creatively about problems and how to solve them.
Data Model: Working in pairs, you will identify your own set of data and construct an entity-relationship/relational model (30%). You will keep a journal of your modeling (10%) and write evaluations of both your final model (30%) and the process that you followed to construct it (30%).	35%	To demonstrate your ability to create a model that captures and describes the data of a specific problem domain at the implementation (relational) level, document the process by which the model was created, and critically assess both your model and the process by which it was created.
Peer Assessment: A grade is awarded to you by the person you are buddying. Staff may also have some input into this mark based on their own observations.	5%	To provide peer support and encouragement and to reinforce the essentially cooperative nature of the modeling activity.
Class Contribution: You are expected to contribute in class on the basis of your reading and practical experiences. Working in randomly assigned groups, you will also be asked to prepare a seminar on a relevant topic. These should be imaginative, creative and interesting.	10%	To encourage lively debate on relevant issues and encourage you to try out new ideas and explore new concepts.
3-Hour Open-Book Examination: There will be two sections in the examination. You will be asked to answer one question in Section 1 (30 minutes) and one question from Section 2.	35%	To provide an opportunity for you to reflect on and apply your learning in this course to situations you may well meet in your early working life. The emphasis will be on your understanding, thinking and application of experiences and not on factual learning.

module, students were encouraged to work in pairs as 'buddies' and one assignment had to be completed with the buddy. All other work, except for the examination, could be completed alone, but the majority of students found it beneficial to work cooperatively, either with their buddy or a small group of students.

Module Content

Using *Data Modeling Essentials: Analysis, Design and Innovation* (Simsion, 1994) as a required textbook, the module focused on standard data modeling issues with a practical rather than a theoretical emphasis. The module was arranged around five broad themes set out in Table 2. The first two themes not only provided the context for the module but also introduced and developed the underlying themes of self-appraisal, critical thinking and problem solving. It is these aspects of these themes that are described here.

SELF-AWARENESS

The first theme encouraged students to become aware of the kinds of activities data modeling entails, and how they, and their peers, approach these

Table 2: Data Modeling Techniques – Broad Themes

Data Modeling – the process	Observing the process of modeling Modeling as analysis or design Modeling as a creative activity
Improving modeling behavior	Common errors of novice modelers – 7 habits of highly effective modelers Problem solving and thinking strategies Interacting with users
Advanced modeling techniques	Advanced normalization Generalization and specialization etc
Evaluation and communicating data models	What is a 'good' model – Understanding the semantics of a model Presenting a model to other people Getting user validation
Data modeling and the real world	The time dimension Generic and enterprise models Business constraints Modeling a data warehouse

tasks. Some time was given to discussing the differing requirements of analysis and design, in particular the different kinds of thinking that they require, for example analytical, scientific thinking and intuitive, creative thinking "that reaches out beyond what is now known into what could be" (Kepner, 1996). Having established that data modeling (at least in the way it was approached in the module) is primarily a design activity, some time was spent in exploring ways of encouraging creative thinking. When the module was first offered, this exploration was mainly confined to common techniques such as brainstorming and mind-mapping, but as the module matured this initial discussion of creativity was linked far more closely to problem solving and thinking strategies as described in the next section. The concept of 'problems,' as having agreed formulations but arguable solutions, as opposed to 'puzzles' and 'messes' (Pidd, 1996), was also discussed.

This theme also incorporated the idea of becoming aware of how an individual actually approaches a modeling exercise. While there has been little academic research in this area, the findings of Srinivasan and Te'eni (1995) were discussed, particularly the suggestion that more experienced modelers are less likely to spend concentrated periods of time focusing on low-level detail. One tutorial was given over to an exercise in which students worked in small groups to complete a number of simple modeling tasks. For each task, a different student played the role of observer and recorded in broad terms the approaches that the group was taking. At the end of the session, the groups looked through these observations and discussed the different approaches that had been used and identified the elements of the approach that either helped or hindered the creation of a model. A number of useful observations often resulted from this exercise, including the effect that a very confident and vocal group member could have on a group's success, that time is often wasted because no strategy for developing the solution has been decided, the way in which different individuals may interpret a problem domain and, of course, the difficulty of finding consensus, particularly when there are deadlines looming! Many of the students came away from this tutorial with a much improved understanding of their own modeling behavior, its strengths and its weaknesses, and began to appreciate the value of having a buddy to both exchange viewpoints with and to use as an observer of their own behavior.

Building on these initial explorations, the second theme developed them further by examining, through discussion and examples, a number of common errors made by novice modelers. This work was based primarily on the errors identified by Batra and Antony (1994), particularly those related to 'anchoring' which they describe as the novice modeler's unwillingness to move away from their initial design. Various strategies, based largely on Moody (1996) and De

Bono (1985, 1993) were suggested for overcoming some of these errors, and students were asked to consider their own strengths and weaknesses, and identify ways in which they would address their weaknesses during the course of their data modeling assignment. Many students recognized their problems with anchoring and the difficulty they found in generating alternative designs, and were thus alerted early on in the module to the need to pay attention to the creation of alternative designs. At this point, the first assignment was introduced, usually to the bemusement of a number of the students.

CREATIVE PROBLEM SOLVING

The first assignment was concerned with finding solutions to problems that were deliberately unrelated to data modeling. The students could choose to provide a solution or solutions to one or more of five problems. If all five problems were attempted, then 20 marks were available for each one, and were graded on the understanding of the problem, the seriousness of the attempt(s) to solve it, and the effectiveness and originality of the proposed solutions. If a student decided to solve less than five problems, the remaining marks were distributed evenly among the attempted questions and awarded for creativity. As another variation, a student could choose to provide a new problem as a substitute for providing two solutions.

The problems had been carefully chosen to illustrate various aspects of creative thinking. The students were told at the beginning that the problems did not have any specifically correct answers, although a reasonable and satisfying answer could be provided to all of them. They were also told that the only way that they would find answers was by using their imagination, thinking laterally and working through a number of possible interpretations of the problems.

The first problem asked the student to construct a grammatically correct and meaningful English sentence that had the word 'and' repeated five times with no other words in between; punctuation between the five 'ands' was allowed. The original intention was that the student would need to recognize the difference between 'and' as a part of a sentence and 'and' as a symbol. For example, one expected solution was "…a sign-writer completed a sign saying 'FishandChips.' Unfortunately, he forgot to put a space between 'Fish' and 'and' and 'and' and Chips…"

In practice, the range of solutions offered each year was astonishing and ranged from re-interpretations of the problem, by, for example, placing numbers between the 'ands' as in "The conductor counted us in by saying 'and 1 and 2 and 3 and 4 and…'," to a clever turn of the tables with "what does

putting and and and and and into a sentence have to do with data modeling."
The initial reaction to the problem was usually one of disbelief in the possibility
of any reasonable solution. However, the majority of students found several
possible solutions to the problem and surprised themselves with their ability to
produce several alternatives.

For the second problem, students were given a 'mystery object,' asked to
study the various design features of the object and then to suggest and justify
a possible use for it. It was made clear that the actual use of the object was
irrelevant and that discovering the real purpose was not the point of the
exercise. In order to score well, the students needed to consider each design
aspect of the object – e.g., its bright orange coloring, the small dimples in the
lid, the sprung steel closing mechanism, the half-moon shape – and attempt to
describe the object that it was designed to hold. There was a generally positive
response to this question. The object was difficult to identify and students'
curiosity was immediately engaged, and the range of solutions offered showed
varying degrees of imagination and problem understanding. High on the
popularity list every year are banana or jelly bean container, suggested by the
shape, although some of the more adventurous have included "The Ultimate Pet
Accessory food bowl, toy box and poop-scoop in one" and "a fine example of
an early aboriginal Survival Boomerang." The latter managed to address all the
obvious design elements while ignoring the question of how a primitive
population was able to source orange plastic and steel springs.

The third problem asked students to identify three things that English does
not appear to have a word for but which perhaps it should. Some students
became fascinated with the process of looking for things that they didn't have
a word to describe and several have remarked on the changes they had noticed
in their own perception following this exercise. Probably the most interesting
answer to this solution came from a student who commented, "I confess that at
the beginning I thought this would be easy and straightforward. What started
as an hour or two in the library turned into many hours of interesting discovery
and enlightenment. The library is a mine of information regarding words. They
provided me with dictionaries and encyclopedias of dictionaries…." This
student went on to suggest the word *fempallas* to describe female intuition
based on a detailed etymology of 'feminine' and Athena Pallas, the Greek
goddess of wisdom, by way of 'philosophy,' 'insight' and 'wise.' The student
concluded, "The process of searching, discovering, evaluation, searching again
and the eventual decision of a 'new' word was interesting. It was difficult to
detail the process because an idea or direction would sometimes just appear.
I find it to be a real problem-solving exercise. I am not 100% sure about the
new word but I learned a lot about searching and evaluation of alternatives.

There were two questions that kept reappearing for me: how do you know which path to follow and how do you know when to stop searching? I do not profess to having discovered the answers, but this exercise made me very conscious of the problem-solving process."

The problem that presented most difficulties was the fourth, where students were asked to construct an English sentence, which is correct and meaningful when spoken but which cannot be written down and remain correct. Solutions generally fell into two categories: statements which use slang expressions which are acceptable when spoken but are not grammatically correct when written, and statements which are logically incorrect when written, e.g., I can't write; This page is blank; This sentence is four words long. Occasionally a student will identify a sentence such as, "There are three (twos) in the English language, two, too and to," having realized that without resorting to phonetics, the sentence cannot be written.

In the final question students were asked to make sense of the expression HOCUS + POCUS = PRESTO and this proved popular among those who decided to treat it as a mathematical equation. Many found mathematical solutions of varying degrees of reasonableness while others used it as a metaphor for the idea that the whole is greater than the sum of its parts. On two occasions, students constructed purely graphical solutions by incorporating the shapes of the letters into a drawing or a painting.

The overall purpose of the assignment gradually became apparent to most of the students as the module progressed. While their initial reaction was often confusion and exasperation, occasionally accompanied by hostility and fear, as they entered into the spirit of discovery, many commented on how they continued to work on and think about the problems long after the assignment had been handed in and graded. Others shared the problems with family or roommates and were surprised by the very different approaches that other people took to solving the same problems. Following the grading of the assignment, a two-hour tutorial session was spent talking about the range of solutions that the class had provided. For many students this was one of the most enlightening aspects of the exercise as they became aware of the variety of solutions that their classmates created. However, it was a question in the final examination, which asked them to comment on what they saw as the purpose of this assignment, which provided some of the most interesting insights into the learning process that had taken place in the preceding three months. Although the question was optional, almost all the students chose to answer it and it was gratifying to see them reflect on their growing awareness of the legitimate existence of differing viewpoints, the possibility of different solutions and their own creativity ability. In addition, most of the students had no difficulty in relating the importance of this to their data modeling (or other design) work. Almost all of the students came to appreciate the limitations in looking only for a 'right' answer, and a significant number of them commented on their increased self-confidence when tackling design problems.

CRITICAL SELF-APPRAISAL

In general, textbooks on data modeling concentrate on the process of constructing a data model and pay little attention to the means by which a data model can be 'read.' One useful exercise is to ask students to exchange a model that they have been working on with another student and ask them to describe what the model appears to be 'saying.' While they are used to interpreting the classic and simplified textbook models, it is often the first time that they have been asked to interpret a complex, incomplete and possibly misleading model. Both sets of students, those whose work is being read and those doing the reading, comment on both the difficulty and the value of this exercise. A technique that can assist them in this work, NaLER (Natural Language for Entity-Relationship Models) by Atkins and Patrick (2000), is taught to students in preference to SERFER (Batra and Sein, 1994) once they have experienced the problem described above. Although the technique can be rather tedious to undertake, it does provide a means for students to appraise the meanings inherent in their own models as they are being constructed. This helps to address the problem of 'outcome-irrelevant learning' noted by Batra and Antony (1994). The technique is straightforward to learn, and some students have remarked that it provides a structure for which they had intuitively recognized a need. Once the NaLER technique had been introduced, the exercise of 'reading' another's model was repeated – this time with the draft version that has been constructed for the second assignment. Students were generally enthusiastic about completing this, partly because they could see the value in having someone else 'proof-read' their model and partly because it offered a sneak preview of what other groups were doing. It was always difficult to restrain students from trying to correct errors that they believed they had identified in another's model. However, when they received back their own set of sentences, they were almost always surprised by the amount of correction their own model required!

The second assignment required the students, working in pairs, to identify a suitable set of data to model, preferably from a real environment, and it was made clear that the primary objective of the assignment was to provide an opportunity to explore, investigate and experiment with ideas and methods of modeling. The assignment description stated that 'how much you learn is more important than the quality of the final model that you produce...To do well in this assignment, it is not sufficient to just produce a model that solves a problem...you will be expected to be able to describe both the faults in your model and in the process by which you created it...." The grading scheme reflected this emphasis by giving 30% of the marks to the model itself and 60% to a critical appraisal of the model and the process (see Table 1).

This scheme was initially driven by two concerns: the difficulty of both fully understanding and objectively assessing more than 20 models from a wide variety of problem domains, and from the belief that once in the workplace, the students would be expected to be aware of the limitations of their own work. By placing significant emphasis on the evaluations, both these concerns were addressed. When grading the model, the marker was not required to fully understand, often with minimal contextual documentation, the problem domain addressed by the model, but could focus on the correct use of constructs, notation and normalization. It thus became the responsibility of the student to describe both the strong and the weak aspects of the model, to identify where further work was required and to, in effect, audit the process that had been followed to create it. This proved to be beneficial to both the students and to the marker, and resulted in any number of very fruitful discussions between them on the reasons for, or the implications of, specific design decisions.

THE CHALLENGE

As an instructor of students who will make their IT careers in the 21st century, one thing is certain - the technologies, the methodologies and the applications that they will learn, use and implement in the course of their careers are likely to be more complex, more sophisticated and very different from those they learned as students. However, the ability to identify, analyze and solve problems, the need to be innovative and creative in their thinking and the ability to step back from their work and appraise it critically, are skills that they will rely on, develop and use throughout their working lives (Sawyer, 1999). As educators we have a responsibility both to our students and to their future employers to provide a sound framework on which these skills can be founded. This is a challenge that we have found difficult to meet in the 20th century, but one which we cannot afford to avoid, for it is these skills, more than any others, which will ensure that today's students become competent and expert practitioners and academics. If we fail, we are severely limiting their chances of successfully meeting the challenges that they will face in realizing the potential of Information Technology in the century ahead.

REFERENCES

Atkins, C. F. & Patrick, J. D. (2000). NaLER: A natural language method for interpreting entity-relationship models. *Campus-Wide Information Systems, 17*(3), 85-93. Available online at www.emerald-library.com.

Batra, D. & Antony, S.R. (1994). Novice errors in conceptual database design. *European Journal of Information Systems, 3*(1), 57-69.

Batra, D., & Sein, M. K. (1994). Improving conceptual database design through feedback. *International Journal of Human-Computer Studies, 40*, 653-676.

Batra, D., & Zanakis, S.H. (1994). A conceptual database design approach based on rules and heuristics. *European Journal of Information Systems, 3*(3), 228-239.

Davis, G. B., Gorgone, J. T., Couger, J. D., Feinstein D. L., and Longenecker, H. E. (1997). *IS'97 Model Curriculum and Guidelines for Undergraduate Degree Programs in Information Systems*. Association of Information Technology Professionals.

de Bono, E. (1985). *Six Thinking Hats*. London: Penguin Books Ltd.

de Bono, E. (1993). *Parallel Thinking*. London: Penguin Books Ltd.

de Carteret, C., and Vidgen, R. (1995). *Data Modeling for Information Systems*. London: Pitman Publishing.

Eaglestone, B., and Ridley, M. (1998). *Object Databases: An Introduction*. London: McGraw-Hill.

Hitchman, S. (1995). Practitioner perceptions on the use of some semantic concepts in the entity-relationship model. *European Journal of Information Systems*, 4, 31-40.

Kepner, C. H. (1996). Calling all thinkers. *HR Focus, 73*(10), 3.

Moody, D. (1996). The seven habits of highly effective data modelers. *Database Programming and Design*, October, 57-64.

Pidd, M. (1996). *Tools for Thinking – Modelling in Management Science*. Chichester, John Wiley & Sons.

Pletch, A. (1989). Conceptual modeling in the classroom. *SIGMOD RECORD*, 18(1), 74-80.

Postman, N., & Weingartner, C. (1969). *Teaching as a Subversive Activity*. New York, Dell Publishing.

Satzinger, J. W., Jackson, R. B., & Burd, S. D. (2000). *Systems Analysis and Design in a Changing World*. Cambridge, MA: Thompson Learning.

Sawyer, D. (1999). *Getting it Right: Avoiding the High Cost of Wrong Decisions*. Boca Raton, FL: St Lucie Press.

Simsion, G. (1994). *Data Modeling Essentials: Analysis, Design and Innovation*. Boston: Van Nostrand Reinhold.

Simsion, G., & Shanks G. (1993). *Choosing Entity Types: A Study of 51 Data Modellers*. Department of Information Systems, Monash University, Working Paper Series 17/93. Melbourne, Australia.

Simsion, G., & Witt, G. (2001). *Data Modeling Essentials Analysis, Design and Innovation, 2nd Edition*. Arizona: Coriolis.

Srinivasan, A. & Te'eni, D. (1995). Modeling as constrained problem solving: An empirical study of the data modeling process. *Management Science*, 41(3): 419-434.

Section II

Teaching Techniques
and Pedagogy

Chapter IV

Towards Establishing the Best Ways to Teach and Learn about IT

Chris Cope, Lorraine Staehr and Pat Horan
La Trobe University, Bendigo, Australia

ABSTRACT

In this chapter we report on an ongoing project to improve the ways we teach about IT in an undergraduate degree. Using a relational perspective on learning, we have developed a framework of factors to encourage students to adopt a deep approach to learning about IT. We describe the design, implementation, evaluation and refinement of learning contexts and learning activities based on the framework. Results are encouraging and show a significant positive effect when compared with a previous study by other researchers involving a different teaching and learning context.

INTRODUCTION

A challenge for IT education in the 21st century is to find ways to produce graduates who meet IT employers' expectations. Recent research indicates that employers are not satisfied with the understanding and skills of new IT graduates, claiming that curricula are unrealistic and unsuited to business needs (Hawforth & Van Wetering, 1994; Misic & Russo, 1996; Roth & Ducloss, 1995; Trauth, Farwell & Lee, 1993).

Copyright © 2002, Idea Group Publishing.

So what is wrong with current IT undergraduate curricula? Is there a problem with the content, as suggested by IT employers? Our review of the IT higher education research literature identified a strong focus on curricular content. Widely available model curricula for IT courses have existed for many years and have been regularly updated and published in the literature (for example Cohen, 2000; Couger et al., 1995; Davis et al., 1997; Longenecker et al., 1995; Mulder & van Weert, 2000). The design of these model curricula incorporates direct input from industry and the research literature on the skills and knowledge requirements of IT professionals (Athey & Wickham, 1995; Doke & Williams, 1999; Lee et al., 1995; Richards & Pelley, 1994; Richards et al., 1998). It is unlikely that the problem lies with the content of IT curricula.

Could the problem lie in 'how' we teach our curricula and, consequently, 'how' our students learn about IT? Are we teaching about IT in the best ways possible? If our strategies for teaching about IT are inappropriate, then it is unlikely that students will be gaining the understanding and skills required by employers. In reviewing the literature on IT teaching, we found reports on implementations of a number of new strategies for teaching about IT. However, these strategies were not, in general, underpinned by any contemporary, research-based view on teaching and learning and have not been evaluated in any structured way. Therefore, the problem of employers' concerns with the skills and understanding of our graduates could lie in 'how' we teach about IT.

In this chapter we report our contribution to overcoming employers' dissatisfaction with IT graduates. We have begun the long process of establishing the best ways to teach and learn about IT. A number of steps have been taken:

1.   We have sought a contemporary, research-based view on student learning which would adequately underpin teaching and learning about IT in the 21st century and should produce graduates with the qualities required by IT employers. We have chosen a relational perspective on student learning (Biggs, 1999; Marton & Booth, 1997; Prosser & Trigwell, 1999; Ramsden, 1987, 1988). From this perspective the best ways to learn about a discipline are to use the deep learning approaches known to lead to high quality learning outcomes. The best ways to teach are to identify and implement in the learning contexts and learning activities the discipline-specific factors likely to encourage students to use deep learning approaches. We describe in detail and justify a relational perspective later in the chapter.

2.   We have reviewed the IT education research literature for insights into the best ways to teach and learn about IT. Our review sought to identify the IT-specific learning context factors and learning activities likely to encourage deep learning approaches. In general we found little attention to the area within the IT education literature.

3.   The next step was an iterative one. Based on a relational perspective on
     learning and, to a limited extent, the IT higher education research literature, we
     have identified and implemented learning context factors and learning activities
     we believe likely to encourage deep learning approaches in IT students.
     Importantly, we have evaluated in a structured way the learning approaches
     students adopted in response to our learning contexts. This has led to an
     ongoing process of modification and evaluation of our learning contexts and
     learning activities.

In this chapter we report our progress to date in attempting to establish the best
ways to teach and learn about IT. While we do not have definitive answers as yet,
we believe our progress is worth reporting. While some of the learning context
factors and learning activities we have identified are not new, the way in which we
designed, justified, implemented and evaluated our teaching and our students'
learning from within a relational perspective on student learning is. We hope that the
chapter will provide fresh insights into teaching about IT and encourage other IT
academics to assist with establishing the best ways to teach and learn about IT. Only
through the use of research-based, considered, evaluated teaching can we hope to
produce graduates of a quality acceptable to IT employers.

BACKGROUND

We believe that a problem confronting IT education at the beginning of the
21st century concerns 'how' the content of our curricula is being taught. To
begin to address this problem, we have reviewed the IT education literature
looking for insights into the best ways to teach and learn about IT. Before
reporting this review, we need to establish what we mean by 'best ways to teach
and learn about IT.' According to Biggs (1999) efforts to improve teaching and
learning require reflection on current practices based on an explicit theory of
learning and teaching. Our explicit theory is a relational perspective on student
learning and teaching, which we now describe. Later in the section we use a
relational perspective to review some of the important teaching strategies and
learning activities proposed in the IT education research literature.

A Relational Perspective on Student Learning

A relational perspective is a contemporary, research-based view of student
learning in higher education, first proposed by Ramsden (1987, 1988), but used in this
chapter to also unify the ideas of other educational researchers, principally
Biggs (1999), Marton and Booth (1997) and Prosser and Trigwell (1999). The

relational perspective research focused on identifying and understanding students' perceptions of their own learning experiences as the impetus for improved teaching and learning. In doing this, the research contrasted with earlier and concurrent studies which focused on external observers' and teachers' perceptions of students' learning experiences, students' behaviours and students' cognitive structures. The relational perspective research identified the nature of and systematic relationships between students' perceptions of their own learning situations, approaches to learning and quality of learning outcomes. We have summarised what we believe to be the essential aspects of a relational perspective on student learning in Figure 1.

Figure 1 shows that the quality of a student's learning outcomes is related to a learning approach which, in turn, is related to the student's perception of their learning situation. A student's perception of their learning situation is influenced by their previous learning experiences and a complexity of contextual factors. We will now describe in more detail the various aspects of the relational perspective illustrated in Figure 1, moving from right to left.

Quality of Learning Outcomes

Many qualitative studies have found that a particular concept, topic or course area can be conceptualised in a limited number of different ways (for

Figure 1: A Relational Perspective on Student Learning (arrows represent empirical relationships)

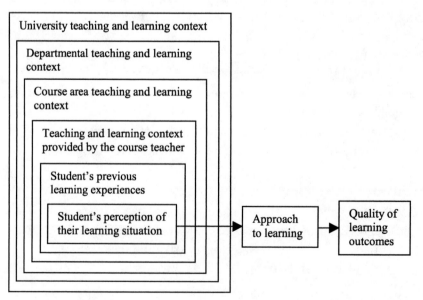

example Crawford et al., 1994; Johansson, Marton & Svensson, 1985; Lybeck et al., 1988; Prosser & Millar, 1989). Of considerable importance, this research has presented the different conceptions in a hierarchy. Higher quality learning outcomes were those conceptions higher in the hierarchy that represented a more complex and complete understanding. Higher quality learning outcomes were found to build on and be inclusive of lower quality learning outcomes. This finding contrasts with some constructivist and cognitive perspectives on learning in which the learning process involves higher level understandings completely replacing lower levels of understanding. These cognitive and constructivist views find it difficult to explain empirical findings that the same student can use different levels of understanding of the same phenomenon in different contexts (Marton, 1998).

Approach to Learning

One of the most important findings of research into student learning from a number of different research perspectives, and central to a relational perspective, is that students' learning approaches can be categorized broadly as either surface or deep. There have been many attempts to generalize the description of surface and deep learning approaches, for example those of Biggs (1999), Marton and Säljö (1976, 1997) and Ramsden (1988, 1992). A comprehensive description is given by Prosser and Trigwell (1999, p.3):

Surface Approach

… students see tasks as external impositions and they have the intention to cope with these requirements. They are instrumentally or pragmatically motivated and seek to meet the demands of the task with minimum effort. They adopt strategies which include focus on unrelated parts of the task; separate treatment of related parts (such as on principles and examples) a focus on what are seen as essentials (factual data and their symbolic representations), the reproduction of the essentials as accurately as possible, and rote memorising information for assessment purposes rather than for understanding. Overall they would appear to be involved in study without reflection on purpose or strategy, with the focus of that study being on the words, the text or the formulae.

Deep Approach

… students aim to understand ideas and seeks meaning. They have an intrinsic interest in the task and an expectation of enjoyment in carrying it out. They adopt strategies that help satisfy their curiosity, such as making the task coherent with their own experience; relating and distinguishing evidence and argument; looking for patterns and underlying principles;

integrating the task with existing awareness; seeing the parts of a task as making up the whole; theorizing about it; forming hypotheses; and relating what they understand from other parts of the same subject, and from different subjects. Overall they have a focus on the meaning in the argument, the message, or the relationships, but they are aware that the meanings are carried by the words, the text or the formulae.

Of great significance to this chapter, a deep approach seeks meaning in a learning task by trying to relate the task to other tasks and/or existing understanding and/or personal experience. A deep approach focuses on developing a cohesive whole by seeking relationships between the various tasks. These relationships are considered to be vital components of an understanding (Entwistle, 1998).

It is important to note that deep and surface learning approaches allow for different preferred learning styles. It is not so much the nature of the way students go about their learning that distinguishes deep and surface learning approaches, but the difference in the intention with which the learning styles are applied. Deep approaches have an intention to understand, surface approaches do not.

Relation Between Approach to Learning and Quality of Learning Outcome

Logically it would seem that high quality learning outcomes involving a deep understanding are unlikely to occur unless a deep learning approach is taken. Only deep learning approaches seek understanding by looking for links between aspects of content. Many empirical studies have confirmed this contention (for example, Cope, 2000; Crawford et al., 1994; Hazel, Prosser & Trigwell, 1996; Marton & Säljö, 1976; Prosser & Millar, 1989).

Students' Perceptions of Their Learning Situation

As deep learning approaches are the vital stepping stone to high quality learning outcomes, it would seem reasonable to investigate how to get students to use deep learning approaches. It has been proposed that learning approach is a characteristic of a student. In fact, students have been found to vary the learning approach they use (Biggs, 1993). Direct instruction in deep learning approaches has been tried, but has been found to lead to poor quality learning outcomes (Marton & Booth, 1997). It seems that instruction in learning techniques leads to a focus on the strategies of learning rather than the intention. A focus on an intention to understand is at the core of a deep learning approach. So how can we encourage students to adopt deep learning approaches? The research has identified certain factors in learning contexts and learning activities which influence the learning approaches students adopt. To understand these factors we firstly need to consider the concept of a student's perception of his/her learning situation.

Many studies of students' perceptions of learning contexts have shown that different students will perceive the same context in different ways (for example, Entwistle & Ramsden, 1983; Prosser & Millar, 1989; Ramsden, 1979). This is not surprising given the complexity of a learning context and the variation among students. When a student sits in a lecture, tutorial, workshop, etc., they are immersed in a context which is far more complex than their immediate surrounds (see Figure 1). Indeed, Biggs (1999, p.25) proposes five critical components of a learning context. These are:

1.   The curriculum being taught
2.   The teaching methods used
3.   The assessment methods used
4.   The environment created in the interaction between the teacher and the students
5.   The institutional (university) climate

In addition to these components, Entwistle and Ramsden (1983) propose that the departmental context is important, and Ramsden (1997) also includes the course area context.

The myriad, complex components of a learning context appear to be filtered by students' prior learning experiences so that only a limited number are perceived by a student at any one time (Prosser & Trigwell, 1999). As students in a particular learning context are likely to vary greatly with regard to previous learning experiences, they will perceive the learning context differently. The interaction between a student's prior learning experiences and a learning context has been called a learning situation by Prosser and Trigwell (1999). Components of a learning context which have been found empirically to be vital parts of a student's perceptions of their learning situation include the quality of the teaching, the clarity of the curriculum goals, the appropriateness of the workload, the nature of the assessment and the amount of choice in learning (Entwistle & Ramsden, 1983). As explained in the next section, it is a student's perception of their individual learning situation which influences the learning approach they will use.

Relation Between Perceptions of Learning Situation and Approaches to Learning

The approach to learning students adopt has been found to be related to perception of learning situation, that is the aspects of the learning context perceived by students as a result of their previous learning experiences (Prosser & Trigwell, 1999; Trigwell & Prosser, 1991; Ramsden et al., 1997). In studies involving many thousands of students across many different subjects, deep learning approaches have been associated with perceptions of good teaching, clear goals and independence in learning. Surface learning approaches have been associated with perceptions of too high a workload and inappropriate assessment, that is, assessment which is perceived to require rote learning.

Returning to our earlier intention to establish what we mean by 'best teach and learn about IT,' part of the answer is now apparent. To best learn about a discipline area like IT means applying a deep learning approach. To best teach about a discipline area like IT means to provide a learning context and learning activities which students are likely to perceive as requiring deep approaches to learning. While the general nature of these learning contexts and learning activities are known, it is significant that the specific details have been shown by research to be discipline dependent (Prosser & Trigwell, 1999). That is, the learning context factors and learning activities found to encourage a deep approach to learning about physics are similar in general but different in detail from the factors likely to encourage a deep approach to learning about humanities or IT, for instance. To establish the best ways to teach about IT, then, we need to identify the IT-specific learning context factors and learning activities which, if perceived by students, will encourage deep learning approaches. To begin this task we turned to the IT education research literature.

A Review of the IT Education Research Literature

We reviewed the IT education research literature hoping to learn about the specific learning context factors and learning activities likely to encourage IT students to adopt deep learning approaches. We found the literature to be limited and unstructured. There was little mention of learning context factors. The focus was on new teaching strategies and, to a lesser degree, learning activities. The impact on students' approaches to learning was rarely mentioned. It was possible, however, to make some limited inferences about the learning contexts intended by the teaching strategies and learning activities.

New teaching strategies were proposed, in general, in response to criticism of the lecture/tutorial strategy considered to be predominant in an IT education context (Avison, 1991). From a general learning perspective, the lecture/tutorial model has been shown by Jackson and Prosser (1989) to be effective in transferring knowledge from a lecturer to students but ineffective in promoting conceptual understanding. From an IT education perspective, the limitations of teaching about IT to undergraduate students using a lecture/tutorial approach are well documented (for example, Cope & Horan, 1994; Fritz, 1987; Little & Margetson, 1989; Mantelaers & Creusen, 1990; Mitri, 1993; Mowete, 1993; Nunamaker, Cougar & Davis, 1982; Osborne, 1992; Schuldt, 1991; Yaverbaum, 1987). The concepts, skills and techniques needed by an IT professional working in a business environment cannot be learned simply through transfer of knowledge in a lecture/tutorial situation, as students do not develop any idea of how to apply the concepts and skills. Academically successful students have been found to pass exams by memorizing the knowledge and mechanically practicing the skills with little or no understanding or mastery of concepts.

In recognition of the problems with a lecture/tutorial teaching strategy, a number of alternative teaching strategies have been proposed in the IT education literature. In general, these strategies were intended to encourage active, cooperative, experiential learning. Some significant examples are now given.

Active participation by students in IT subjects has been promoted as a contrast to the passive approaches encouraged by the conventional lecture/tutorial model. A preference by students for active rather than passive participation was demonstrated by Lu (1994). Students perceived active participation as a more effective way to learn and made more effort when taught in this way. Active participation was encouraged by Wagner (1997) through using the World Wide Web (WWW) as a discovery-based learning tool. Students were encouraged to learn for themselves rather than being told the answers by the lecturer. Learning then becomes an active process of "modifying one's mental framework or knowledge chunks to allow them to comprehend a broader range of life experiences rather than the memorization of discrete facts" (p.76). This learning process is compatible with a deep approach to learning.

An activity that can be used to extend the advantages of active participation in IT subjects through the incorporation of cooperative learning is role plays (Cope & Horan, 1996; Mowete, 1993, Sullivan, 1993). Cope and Horan, for instance, used role plays in introductory IS classes as an adjunct to a lecture/tutorial teaching approach. The role plays involved a re-creation of a real-life scenario, with the students role playing an IS developer and an actor role playing the client. The students, working in small groups, were required to interview the client three times, once in a large group lecture situation and twice in small-group tutorial situations. The interviews were a means of determining the client's requirements for an IS. Cope and Horan evaluated the impact of the role play activity by investigating students' perceptions of the learning situation in which the role play activity took place. The evaluation sought to determine if students perceived the need to apply a deep learning approach to the role play activity. The role plays were found to be more effective than a lecture/tutorial approach in encouraging students to use deep learning approaches.

An activity intended to extend the advantages of active, cooperative learning by making learning more experiential is the use of small-group, long-term, real-world or simulated projects (for example, Crow & Rariden, 1992; Little & Margetson, 1989; Mukherjee & Cox, 1998). Projects generally require students working cooperatively in small teams to research an information storage and retrieval problem for a client in a business environment. The project includes the design, construction, documentation and implementation of a solution. In general projects are of an extended length, compared to more usual learning tasks such as assignments. IT projects are experiential in that they are "hands on" and provide students with an experience of a business context. An advantage of projects claimed by Crow and Rariden (1992 p.53) is that "Students will learn better if involved in an active and real mode."

Two other significant examples in the literature described learning activities that focused on principles compatible with deep learning approaches. These principles were the seeking of the meaning of the content being taught and the necessity for students to be aware of their developing understanding of the content and approaches to learning about IT.

The first example by Godfrey (1995) used Participative Course Design (PCD) to encourage holistic learning approaches in IT students with a focus on understanding concepts and principles. PCD is a collaborative instructor-student activity which establishes shared expectations for learning goals and learning processes early in a subject. While Godfrey does not describe holistic learning approaches in any detail, Svensson (1984) describes them as trying to form a cohesive whole. This is consistent with a deep approach to learning where students try to relate the different aspects of the content to one another.

The second example by Thomson (1994) describes the use of the principles of Total Quality Management (TQM) in IS subject design and implementation. Applying TQM to learning about IS involved considering students as clients and focusing on the learning process and its continued improvement. Students were required to view the classes as learning experiences, and design and implement their own way of learning about their choice from a number of topics. Regular meetings of the group of students exposed students to variation in understanding and the learning process, leading to continual improvement in both the learning process and understanding of the content. The use of TQM in this way is consistent with the ideas of Marton and Booth (1996, 1997) and Cope (2000) that the intention to understand, so critical to a deep learning approach, is best encouraged by having students reflect on the differences between their own and other students' levels of understanding and approaches to learning.

Considering the relevant contributions from the IT education research literature, what can be inferred about the IT-specific learning context factors and learning activities likely to encourage deep learning approaches? Little is known about the specific learning context factors appropriate for IT learning contexts, although learning activities should involve IT students trying to seek a cohesive whole with regard to the content. To achieve this aim, students need to reflect on both the content and approaches to learning about the content. Experiential learning activities are needed in which learning is active rather than passive and in small groups where students have to co-operate in the learning process.

Considering the limited input from the IT education research literature, from a relational perspective we must conclude that the best ways to teach and learn specifically about IT have not been established. We cannot rule out inadequate teaching as the cause of employers' dissatisfaction with our graduates. At the beginning of the 21st century, this is not a good situation for IT education.

To move towards establishing the best ways to teach and learn about IT, we have, over a period of time, used the research-based knowledge inherent in a relational perspective on learning to change the nature of our learning contexts and learning activities. On the basis of an iterative process of structured research, we have designed and implemented these learning contexts and activities, evaluated the resultant learning approaches used by students, and refined the learning contexts and activities. In the next section we describe the framework we used to give structure to our research.

A Framework for Action

The purpose of changing our learning contexts and learning activities was to encourage IT students to use deep learning approaches. Only through using deep learning approaches are IT undergraduates likely to develop the understanding and skills required by employers. As a benchmark for the best ways to learn about IT, and against which to design our changes and to evaluate their impact, we have used the comprehensive description of surface and deep learning approaches given by Prosser and Trigwell (1999, p.3), presented earlier in the chapter.

To encourage our students to use deep learning approaches as described by Prosser and Trigwell (1999), we have constructed a framework of factors around which to design our learning contexts and activities. The factors were derived principally from the general relational perspective literature (Biggs, 1999; Gibbs, 1992; Martin et al., in press; Marton & Booth, 1996, 1997; Ramsden, 1988, 1992) with some input from our review of the IT education research literature. The factors have been established either empirically or logically as encouraging deep learning approaches. The framework is divided into factors relating to the learning context and those relating to learning activities.

The factors included in the framework guided the design of our learning contexts and the learning activities. The practical implementation of the framework is described below.

Table 1: Framework of Factors Likely to Encourage Deep Learning Approaches

1.	*The learning context* – The following factors are important to achieving a learning context likely to encourage deep learning approaches:
a.	The course is well organized and has clear goals
b.	Teaching approaches support the explicit aims and objectives of the course
c.	The student has responsibility for his/her own learning, including some control over the content and approach to learning
d.	The workload is manageable
e.	The student is given help in learning within the context of the subject matter
f.	The teaching makes the structure of the individual topics, and the course as a whole, explicit
g.	Assessment tasks require the demonstration of conceptual understanding
h.	Teaching and assessment methods foster active and long-term engagement with learning tasks
i.	The teaching identifies and builds on what students already know
j.	A supportive classroom environment is provided where students feel comfortable to openly discuss their conceptions and learning approaches
k.	The emphasis is on depth of learning rather than breadth of coverage
l.	The teaching is stimulating and demonstrates the lecturer's personal commitment to the subject matter and stresses its meaning and relevance to the students
m.	Feedback is appropriate and timely
2.	*The nature of learning activities* – The following factors are important in the design of learning activities. Learning activities should:
a.	Be active and experiential
b.	Tackle real-world problems that are compatible with the experiences of the students
c.	Encourage students to relate the learning to situations outside the educational context
d.	Encourage students to reflect on the content *and* the learning process
e.	Use group interaction to expose students to variation in the ways other students understand the content and approach their learning

IMPLEMENTATION

We have implemented the framework in the learning contexts and activities we use in courses in the information systems stream of the Bachelor's of Computing taught in the Department of IT at La Trobe University, Bendigo. We have evaluated the impact of our learning contexts and activities over a number of years and consequently refined them. A brief description of the latest implementation follows. In the description, the figures in brackets (for example 1b) refer to the specific learning context factors and learning factors which make up our framework for action, described in Table 1. Full detail of our learning contexts and learning activities can be found in Cope, Horan and Garner (1997), Horan (1996, 2000), Staehr (1999) and Staehr and Cope (1999).

Our current teaching of IT includes the following features.

Learning Objectives

We specify learning objectives for each course and for each topic in a course. Following Ramsden (1997), the objectives clearly specify learning outcomes or processes which should result from studying the course or topic. The objectives are published on the Web and are frequently referred to by the lecturer during classes (1b, 1f). The objectives present students with the course goals (1a) and enable them to evaluate their progress throughout the semester (2d).

Workshops

We have found that most of the factors in the framework are implemented more effectively in workshops than lectures. Consequently we have restructured the delivery of our courses. For example, in one subject, a two-hour workshop and a one-hour tutorial each week have replaced two one-hour lectures and two one-hour tutorials. We have reduced the total number of contact hours per week as we expect our students to spend at least an hour a week on the reading material they are given (1d). The reading material comes from a book of readings we prepare each year as an alternative to prescribing a textbook.

We use the Web to indicate which topics are to be covered in a particular workshop, to provide an introduction to each topic and to indicate the readings which are appropriate. Prior to attending a workshop students are expected to identify the topic that is relevant, to read the appropriate readings and to make their own notes on the topic. Not all topics relate to the information systems discipline, some are specifically included to help the students learn, for example, note taking, concept mapping and rich pictures (1e).

A typical workshop contains 25–30 students and is structured in the following way. During the first 20 minutes or so, the lecturer gives an overview of the topic and relates the topic to the course as a whole (1f). Any questions arising from the readings are dealt with by a general discussion. The remainder of the workshop involves students actively working on experiential problems in small groups (typically three or four students) (2a, 2e). These learning activities explore the topic for the workshop. The activities are centered around a case study designed to be compatible with the real-world experiences of the students (2b, 2c). Student groups are asked to report their solutions to the problems to the rest of the students in the class, thus exposing students to variation in levels of understanding (2e). This workshop format gives staff the means and time to consult with the small student groups to discover the stage of learning reached by individual students and to demonstrate limitations in existing student conceptions (1i, 1j).

In the middle and at the end of semester, workshops are allocated which do not introduce new topics but allow students to consolidate the topics already covered and to catch up with assignment work (1d, 1k, 2d). In these workshops, learning activities designed to integrate topics are used. For more detail on our use of workshops, see Staehr and Cope (1999).

Rich Pictures

We have used Rich Pictures (RPs) as the cornerstone of our learning contexts. RPs were originally proposed by Peter Checkland (1981), as part of his Soft Systems Methodology. Checkland used RPs to graphically describe and analyze IS contexts. Important features of rich pictures are that they are flexible and non-prescriptive, can be used and interpreted by anyone, and can be used to display concrete and abstract factors. Our students use RPs, individually (1c) and in groups (2e), both as a standard IS documentation tool (2c) and, importantly, as a way of developing, reflecting on and recording both their understandings of our topics and courses and the learning process (2d). RPs also provide a medium for discussion with and feedback from the teacher (1j, 1m). For a detailed description of our use of RPs, see Horan (2000).

Case Studies

We use case studies of typical business applications as a means to integrate the content of our Courses (1f). A case study is introduced at the beginning of the semester, and the weekly topics over the semester are all introduced and developed within the context of the case study. This provides continuity as different topics are progressively illustrated in the same context (1h). The experiential learning activities in the workshops (2a), mainly group-based (2e), are based on the case study, and some of the assessable assignment work involves extending aspects of the case study. Although of necessity simplified for the classroom, the case study links the topics of the course to the real world (2b, 2c). For more detail on our use of case studies, see Cope and Horan (1996) and Horan (1996).

Assessment

Each assignment we give to our students lists the learning objectives underlying the design of the assessment (1b). All assignment work is marked, annotated and returned to students promptly so that they can evaluate their progress and seek early assistance if needed (1m). The learning portfolio assignment best illustrates the application of our framework to assessment. Students are asked to choose any two topics covered in the course so far and to demonstrate their understanding of those topics using two different methods. We place strong emphasis on the importance of demonstrating a deep understanding of the material rather than presentation of isolated facts (1g). Some methods are suggested, for example, letter to a friend, a workshop plan or a rich picture, but students are free to come up with their own method (1c). Students are given the opportunity to have a draft of one of their topics commented on by staff well before the assignment is finally handed in for marking (1m).

In summary, the implementation of our framework (see Table 1) was intended to modify the learning contexts and learning activities we provide for our students. The framework was used in such a way that students' previous learning experiences and perceptions of their learning situation could be perceived by the teacher and made explicit to the students. It was incorporated into the learning context, including materials and tasks made available to the students, both within and outside formal classes. All of the techniques described reinforce one another by exhibiting characteristics likely to encourage a deep learning approach.

EVALUATION

Structured evaluations of new teaching strategies are essential to establishing the effectiveness of different ways to teach and learn about IT, yet are not commonly reported in the literature. We have endeavored to evaluate quantitatively and qualitatively the impact of the changes to our learning contexts and learning activities. A description of one of these evaluations and the outcomes is now given. For further examples of our evaluations, see Cope (1997, 2000), Cope, Horan and Garner (1997), Cope and Horan (1996, 1998) and Staehr and Cope (1999).

Quantitative Evaluation

We have centered the quantitative evaluations of our learning context and learning activity changes on the learning approaches used by students taking the second-year course, Information Systems Development. From a relational perspective on learning, the aim of any changes to learning contexts or activities should be to encourage more students to use deep learning approaches. The impact of any changes can be evaluated by comparing the numbers of students adopting surface and deep approaches before and after the changes.

In Information Systems Development we changed from a lecture/tutorial to a workshop/tutorial teaching strategy in 1998. The workshop/tutorial strategy attempted to implement the learning contexts and learning activities we described in the previous section. At the end of each implementation of Information Systems Development, from 1997 (before the change) through 2000, students completed a questionnaire (Appendices A and B). The questionnaire contained a number of Likert scale questions from the Study Process Questionnaire (Biggs, 1987) and short, open-ended items which investigated the students' approaches to learning. Using the descriptions of surface and deep learning approaches of Prosser and Trigwell (1999), two researchers independently categorized each questionnaire as representative of a surface, surface/deep or deep learning approach, following Prosser and Millar (1989). A surface/deep learning approach demonstrated some, but not all, elements of a deep approach. The researchers then compared categorizations and discussed differences until agreement was reached. The results are shown in Table 2.

To obtain an indication of the consistency of academic ability over the study period, the mean Tertiary Entrance Ranks (TER) of students in each year was compared using a single-factor analysis of variance. A TER is the university entrance score of a student in Victoria, Australia, obtained from academic

Table 2: Students Categorized by Learning Approach

Year	\multicolumn{6}{c}{Learning Approach}					

	Surface		Surface/Deep		Deep	
	No. Students	% Students	No. Students	% Students	No. Students	% Students
Year						
1997	45	76.3	8	13.6	6	10.2
1998	39	66.1	14	23.7	6	10.2
1999	45	77.6	8	13.8	5	8.6
2000	35	62.5	17	30.4	4	7.1

performance in the last year of secondary school. The results of the single factor analysis of variance showed no significant differences in TER between years. Therefore, with respect to TER, and by implication academic ability, the profile of students was similar for each year.

The results in Table 2 were analyzed with a Chi-square test for independence using exact significance levels (SPSS exact tests). This test showed no interaction between year and learning approach and therefore indicates similar response profiles across all years. In other words our changes had no statistically significant effect on the number of students using major aspects of deep approaches in their learning.

Table 3 compares our results in the year 2000 (see Table 2) using the workshop/tutorial learning context with a study by Prosser, Walker and Millar (1994) that involved a lecture/tutorial learning context. We chose our results from the year 2000 because by this time, the two lecturers involved were thoroughly familiar with the new learning context. By studying the method used by Prosser, Walker and Millar, we determined that by combining our surface/deep and deep categories, we could make a valid comparison of the results of the two studies using 95% confidence intervals for the percentages in each cohort. The upper 95% confidence limit for students using aspects of a deep approach to their learning is 14.2% in the study by Prosser, Walker and Millar, and our lower limit is higher at 24.8%. Therefore, our teaching strategy resulted in a significantly higher proportion of students using significant aspects of a deep approach to learning in comparison.

Table 3: A Comparison of Two Studies Using Different Learning Contexts (the errors are based on the 95th percentiles of the normal distribution and represent two standard deviations)

Study	Course	No. of students	Learning context	Significant aspects of a deep approach	Surface approach
Prosser, Walker & Millar (1994)	Physics	84	Lecture/ tutorial	8.3% ± 5.9%	91.7% ± 5.9%
Table 1 (2000 results)	Information Systems	56	Workshop/ tutorial	37.5% ± 12.7%	62.5% ± 12.7%

Qualitative Evaluation

Qualitative feedback has been obtained from our students through in-depth interviews, questionnaires and focus groups. In general, students readily perceived the difference between the deep learning approaches required in our subjects and the rote learning surface approaches required by many of their other subjects. Although many found the deep approach more difficult than a surface learning approach, they acknowledged its value.

We discuss the implications of our evaluations in the next section. We also consider how a global research approach is required to most efficiently establish the best ways to teach and learn about IT. This will be an important challenge for IT education in the 21st century.

DISCUSSION AND CONCLUSION

We began this chapter by suggesting that the concerns expressed by employers about the quality of IT graduates may be the result of problems with 'how' we teach and 'how' our students learn about IT. Our review of the IT education research literature and our evaluation results support this proposition. Even after the implementation and refinement of the changes we have made to our learning contexts and learning activities, fewer than half of our students were using significant aspects of the deep learning approaches likely to lead to the understandings and skills required by employers. The majority of our students are not learning about IT in the best ways. Given that recent research by Prosser, Trigwell and Taylor (1994) and Trigwell and Prosser (1996) has demonstrated an empirical relationship between teachers' approaches to teaching and students' approaches to learning, we must also conclude that we continue not to be teaching about IT in the best ways possible.

We have proposed that teaching and learning about IT based on a relational perspective on learning, and the framework of learning context and learning activity factors derived from the literature, should, logically, lead to an improvement in the proportion of students using deep learning approaches and resultant higher quality learning outcomes. What empirical evidence do we have regarding this claim? Our evaluation of the course Information Systems Development found no statistically significant change between 1997 and 2000 in the proportion of students using deep learning approaches in response to our interventions. Should we continue with a relational learning perspective and our framework of learning context and learning activity factors as the foundation for our attempts to establish the best ways to teach and learn about IT? We now present what we believe to be evidence that encourages us to continue our efforts.

Our failure to detect a statistically significant change in the proportion of students using deep learning approaches is not unexpected. One possible cause of this failure may be lack of statistical power due to the relatively small numbers of students in each year. In our IT context we would need a very large change in students' learning approaches to produce a statistically significant change. Given the generally recognized slowness of educational change, we are not discouraged by our results. Indeed we, as teachers, have been learning as we go in implementing our new learning contexts and learning activities. We did not expect immediate and striking results.

Further encouragement for continuing with the thrust of our present research comes from the significant difference we have been able to demonstrate in the comparison of the results of our study in the year 2000 with those of Prosser, Walker and Millar (1994). The results of this comparison are encouraging for two reasons. Firstly, the students in the Prosser, Walker and Millar study were first-year undergraduate students and our students were second-year students. Biggs (1987, p.8) states that "there is a general decline in the deep approach from first to final year for those completing first degrees in both universities and colleges." This decline in the deep approach was shown in Prosser, Walker and Millar's study. At the beginning of the first-year physics course, $26.2 \pm 5.9\%$ of the students surveyed were using significant aspects of deep approaches and at the end of the year there were only $8.3 \pm 5.9\%$. In contrast the students in our study were in the middle of their second year and a higher proportion ($37.5\% \pm 12.7\%$) were using significant aspects of deep approaches. Biggs (1987, p.8) goes on to say that "Those students continuing with postgraduate study, however, show a marked rise in deep approach." The students in Prosser, Walker and Millar's study were from a prestigious, major metropolitan university where students would be encouraged to consider research careers. In contrast, few of our students continue on to postgraduate study. Secondly, students in Prosser, Walker and Millar's study would have had university entrance scores considerably higher than those of our students. The Bendigo campus of La Trobe University is in a rural area and takes in students with a much broader range of university entrance scores than do metropolitan universities. Therefore, we may have been even more successful in improving our students' approaches to learning when these factors are taken into account.

We believe we have successfully begun the task of establishing the best ways to teach and learn about IT. A relational perspective on learning and the framework of learning context and activity factors we developed are worth pursuing as foundations for improvement. Learning contexts and learning activities grounded in this perspective and framework are better ways to teach and learn about IT. So what do we do next?

Insight can be gained from considering why our efforts have not been more successful. Our evaluation method involved investigating how students approached their learning. While providing insight into the overall effectiveness of our interventions, this method does not tell us why the majority of students continue to use surface learning approaches and, hence, how we can improve matters further. From a relational perspective on learning, students using surface learning approaches are not perceiving the learning contexts as we intended. Identifying which factors need further manipulation is critical to continuing our efforts to establishing the best ways to teach and learn about IT.

We have some evidence from the data collected from the course, Information Systems Development, to suggest that students do not perceive their workload as being manageable (1d.). This evidence is described in detail in Staehr and Cope (1999) and comes from further statistical analysis of the Likert scale questions on our measuring instrument (Appendices A and B). This is despite the fact that we have decreased the number of contact hours and content we cover and included two weeks of consolidation where we do not cover new content. We concluded that the problem could well lie in the workload in other courses the students were undertaking concurrently with Information Systems Development and the fact that full-time students are increasingly having to take on part-time work to pay for their education. Our research indicates that we need to begin to try to influence these broader contexts in a structured, evaluated way. We need to talk to other lecturers in our department and at our university and inform them of our commitment to, enjoyment in and strategies for improving our teaching and our students' learning. We need to communicate the encouraging results of our quantitative and qualitative evaluations. We need to negotiate our students' workloads at a department and university level.

We have no evidence of which other factors in our framework are being perceived inappropriately by our students. Research into students' perceptions of our learning contexts is needed to isolate the IT-specific learning context and activity factors we need to manipulate further. Further research will involve using the Course Experience Questionnaire (Ramsden, 1991) which has sub-scales for the factors known to encourage surface and deep approaches. In addition in-depth interviews with small groups of students about perceptions of their learning situations are necessary to provide us with a more complete picture.

Given the myriad of factors which make up the learning contexts we provide for our students, we conclude our chapter with a call to arms. Our research has shown that the task of establishing the best ways to teach and learn about IT is clearly incomplete, is a complex one and is an important challenge for IT education in the 21st century. The task is beyond the efforts of a small research group. A global, collegial approach to IT education research is called for to take into account global variations in culture, history and location of teaching and learning contexts. There is a need on a global scale to design and implement new IT learning contexts and activities as improvements to those already evaluated and reported in the literature. It follows that there is a need to make ourselves aware of what has already been done through reference to the literature and to make relevant, justified changes that are grounded in an accepted perspective on student learning. Reported evaluations of interventions to teaching strategies are critical. The evaluations need to be considered, structured and grounded in theory.

The time to establish the best ways to teach and learn about IT is now. Any present and future view of our world indicates the growing importance of IT and hence IT education. We must actively seek to make our teaching of IT effective. The task is beyond an individual. The challenge is a global one. The global communication means are available in terms of relevant electronic journals and international conferences. The call is for published, structured research grounded in the existing literature and contemporary views of student learning and teaching.

APPENDIX A

Short answer questionnaire items used to investigate approach to learning in IS Development in 1997 (lecture/tutorial strategy)

1.   Describe what **you** do during lectures in IS development.
2.   How does what you do in the lectures assist you to learn?
3.   Describe what **you** do during tutorials.
4.   How does what you do in the tutorials assist you to learn?
5.   Describe how you have studied for the assignments in IS development.
6.   Describe any ways in which the assignments in IS development have assisted you to learn.
7.   Describe any additional study you have done at home that has contributed to your learning in IS development.
8.   How has this study assisted you to learn?
9.   Describe in words or pictures what *'learning'* means to you.

APPENDIX B

Questionnaire items used to investigate approach to learning in IS Development in 1998, 1999 and 2000 (small student group strategy)

Likert scale questions (5 - True all of the time, 4 - True most of the time, 3 - True half of the time, 2 - True little of the time, 1 - True none of the time)

1. The best way for me to understand technical terms in this subject is to memorize definitions.
2. I find I have to concentrate on memorizing a good deal of what we have to learn in this subject.
3. I try to relate ideas in this subject to those in others, wherever possible.
4. I find it helpful to 'map out' a new topic in this subject by seeing how the ideas fit together.
5. In this subject, I generally put a lot of effort into trying to understand things which initially seem difficult.
6. I usually set out to understand thoroughly the meaning of what I am asked to read in this subject.
7. Often I find I have read things in this subject without having a chance to really understand them.
8. In trying to understand new ideas, this subject encouraged me to relate them to real-life situations.
9. I usually don't have time to think about the implications of what I have read in this subject.

Open-ended questions
1. Why do you think your lecturers chose to teach ISD in the way they did (i.e., having workshops rather than lectures and expecting you to read about each topic beforehand)?
2. How have you gone about learning in ISD? Say what you actually did rather than what you think you should have done.
3. Why did you go about learning in this way?
4. Describe what you did in the workshops in Information Systems Development. Say what you actually did rather than what you think you should have done.
5. How did your activities in the workshops contribute to your learning in the subject Information Systems Development?
6. Describe any work related to Information Systems Development that you did outside of class time. Say what you actually did rather than what you think you should have done.
7. How did your work outside of class time contribute to your learning in the subject Information Systems Development?

ACKNOWLEDGMENTS

The teaching initiative in the course, Information Systems Development, was supported financially by the La Trobe University, Bendigo Teaching Committee and by the School of Management, Technology and Environment Research Advisory Group.

Dr. Graeme Byrne, Division of Mathematics, La Trobe University, Bendigo, for assistance with the statistics in the project.

REFERENCES

Athey, S., & Wickham, M. (1995). Required skills for information systems jobs in Australia. *Journal of Computer Information Systems*, 36(2), 60-63.

Avison, D.E. (1991). Action programmes for teaching and researching information systems. *The Australian Computer Journal*, 23(2), 66-72.

Biggs, J. (1987). *Study Process Questionnaire Manual*. Melbourne: Australian Council for Educational Research.

Biggs, J. (1993). From theory to practice: A cognitive systems approach. *Higher Education Research and Development*, 12, 73-85.

Biggs, J. (1999). *Teaching for Quality Learning at University: What the Student Does*. Buckingham: Society for Research into Higher Education & Open University Press.

Checkland, P. (1981). *Systems Thinking, Systems Practice*. Chichester: Wiley.

Cohen, E. (Ed.) (2000). Curriculum Model 2000 of the Information resource management Association and the Data Administration Managers Association. [On-line]. Available: http://gise.org/IRMA-DAMA-2000.pdf.

Cope, C.J. (1997). Learning about information systems: A relational perspective. *Proceedings of HERDSA '97*, Adelaide, July 8-11.

Cope, C.J. (2000). Educationally critical aspects of the experience of learning about the concept of an IS. Unpublished PhD thesis. La Trobe University, Australia. [On-line]. Available: http://ironbark.bendigo.latrobe.edu.au/~cope/cope-thesis.pdf.

Cope, C. J., & Horan, P. (1994). An alternative curriculum for introductory information system students. *Proceedings of the Eleventh Information Systems Education Conference*, Louisville, U.S., October 28-30, pp. 167-173.

Cope, C. J. & Horan, P. (1996). The role played case: An experiential approach to teaching introductory information systems development, *Journal of Information Systems Education*, 8 (2&3), 33-39.

Cope, C. J. & Horan, P. (1998). Toward an understanding of teaching and learning about information systems. *Proceedings of the Third Australasian Computer Science Education Conference*, Brisbane, July 8-10th, pp. 188-197.

Cope, C. J., Horan, P. & Garner, M. (1997). Conceptions of information systems and their use in teaching about IS. *Journal of Informing Science*, 1(1), 8-22.

Couger, J., Davis, G.B., Dologite, D.G., Feinstein, D.L., Gorgone, J.T., Jenkins, A.M., Kasper, G.M., Little, J.C., Longenecker, Jr., H.E., & Valacich, J.S. (1995). IS'95: Guidelines of undergraduate IS curriculum. *MIS Quarterly*, 19, 341-359.

Crawford, K., Gordon, S., Nicholas, J., & Prosser, M. (1994). Conceptions of mathematics and how it is learned: The perspectives of students entering university. *Learning and Instruction*, 4, 331-345.

Crow, G.B., & Rariden, R.L. (1992). Strengthening the computer information systems curriculum through cooperative education. *Journal of Computer Information Systems*, 32(3), 52-55.

Davis, G.B., Gorgone, J.T., Couger, J.D., Feinstein, D.L., & Longenecker, H.E. Jr. (1997). IS'97: Model curriculum and guidelines for undergraduate degree programs in information systems. *Joint Report of the Association for Computing Machinery (ACM), Association for Information Systems (AIS) and Association of Information Technology Professional (AITP)*. Available from ACM and AITP.

Doke, E., & Williams, S. (1999). Knowledge and skill requirements for information systems professionals: An exploratory study. *Journal of IS Education*, 10(1), 10-18.

Entwistle, N. J. (1998). Improving teaching through research on student learning. In Forest, J. F. F., (Ed.), *University Teaching: International Perspectives*. New York: Garland Publishing.

Entwistle, N. J., & Ramsden, P. (1983). *Understanding Student Learning*. London: Croom Helm.

Fritz, J. (1987). A pragmatic approach to systems analysis and design. *SIGCSE Bulletin*, 19(1), 127-131.

Gibbs, G. (1992). *Improving the Quality of Student Learning*, Bristol: Technical and Educational Services.

Godfrey, R.M. (1995). Students as end-users: Participative design of the I/S learning experience. *Journal of Computer Information Systems*, 36(1), 17-22.

Hawforth, D.A., & Van Wetering, F.J. (1994). Determining underlying corporate viewpoints on information systems education curricula. *Journal of Education for Business*, 69, 292-295.

Hazel, E., Prosser, M., & Trigwell, K. (1996). Student learning of biology concepts in different university contexts. *Research and Development in Higher Education*, 19, 323-326.

Horan, P. (1996). Teaching information systems analysis and design using case studies. *Proceedings of PRIISM '96*, Hawaii, January, pp. 39-46.

Horan, P. (2000). Using rich pictures in information systems teaching. *Proceedings of the first International Conference on Systems Thinking in Management*, Geelong, November, pp. 257-262.

Jackson, M.W., & Prosser, M.T. (1989). Less lecturing, more learning, *Studies in Higher Education*, 14(1), 55-68.

Johansson, B., Marton, F., & Svensson, L. (1985). An approach to describing learning as change between qualitatively different conceptions. In Pines, A. L., & West, L. H. T., (Eds.), *Cognitive Structure and Conceptual Change* (pp. 233-258). New York: Academic Press.

Lee, D. M. S., Trauth, E.M., & Farwell, D. (1995). Critical skills and knowledge requirements of IS professionals: A joint academic/industry investigation. *MIS Quarterly*, 19, 313-340.

Little, E., & Margetson, D.B. (1989). A project-based approach to information systems design for undergraduates. *The Australian Computer Journal*, 21(2), 130-136.

Longenecker, H.E. Jr., Feinstein, D. L., Couger, J. D., Davis, G.G., & Gorgone, J.T. (1995). Information Systems '95: A summary of the collaborative IS curriculum specification of the joint DPMA, ACM, AIS task force. *Journal of Information Systems Education*, 6(4), 174-186.

Lu, H. (1994). A preliminary study of student responses to different CIS course teaching strategies. *Journal of Computer Information Systems*, 34(4), 31-36.

Lybeck, L., Marton, F., Stromdahl, H., & Tullberg, A. (1988). The phenomenography of the "Mole Concept" in chemistry. In Ramsden, P., (Ed.), *Improving Learning. New Perspectives* (pp. 81-108). London: Kogan Page.

Mantelaers, P.A.H.M., & Creusen, M.W.F.J. (1990). Teaching information systems design: Mission impossible? In Richardson, J., (Ed.), *Computers in Education: Conference Abstracts. IFIP Fifth World Conference on Computers in Education (WCEC90)*. Australian Council for Computers in Education.

Martin, E., Prosser, M., Trigwell, K., Leuckenhausen, G., & Ramsden, P. (In press). Using phenomenography and metaphor to explore academics' understanding of subject matter and teaching. In Rust, C., (Ed.), *Improving Student Learning*. Oxford: Oxford Centre for Staff Development.

Marton, F. (1998). Towards a theory of quality in higher education. In Dart, B., & Boulton-Lewis, G., (Eds.), *Teaching and Learning in Higher*

Education (pp. 177-200). Camberwell, Victoria, Australia: Australian Council for Educational Research.

Marton, F., & Booth, S. (1996). The learner's experience of learning. In Olson, D. R., & Torrance, N., (Eds.), *The Handbook of Education and Human Development: New Models of Learning, Teaching and Schooling* (pp.534-564). Oxford: Blackwell.

Marton, F., & Booth, S. (1997). *Learning and Awareness*. Mahwah, NJ: Erlbaum.

Marton, F., & Saljo, R. (1976). On qualitative differences in learning. I. Outcome and process. *British Journal of Educational Psychology*, 46, 4-11.

Marton, F., & Saljo, R. (1997). Approaches to learning. In Marton, F., Hounsell, D., & Entwistle, N. J., (Eds.), *The Experience of Learning: Implications for Teaching and Studying in Higher Education* (2nd Ed.) (pp. 39-58). Edinburgh: Scottish Academic Press.

Misic, M., & Russo, N. (1996). Educating systems analysts: A comparison of educators' and practitioners' opinions concerning the relative importance of systems analyst tasks and skills. *Journal of Computer Information Systems*, 36(4), 86-91.

Mitri, M. (1993). The role-play exercise for an introductory computer information systems course. *Interface*, 15(2), 39-43.

Mowete, R.G. (1993). Enhancing conceptual learning in the systems analysis course. *Interface*, 14(4), 2-6.

Mukherjee, A., & Cox, J.L. (1998). Effective use of mastery-based experiential learning in a project course to improve skills in systems analysis and design. *Journal of Computer Information Systems*, 38(4), 46-50.

Mulder, F. & van Weert, T. (2000). Informatics Curriculum Framework 2000 for higher education. Publication of the International Federation for Information processing (IFIP). Available: http://Poe.netlab.csc.villanova.edu/ifip32/ICF2000.htm.

Nunamaker, J.F., Cougar, J.D., & Davis, G.B. (1982). Information systems curriculum recommendations for the 80s. *Communications of the ACM*, 25(11), 781-805.

Osborne, M. (1992). APPGEN: A tool for teaching systems analysis and design. *SIGCSE Bulletin*, 24(1), 259-263.

Prosser, M., & Millar, R. (1989). The how and what of learning physics. *European Journal of Psychology in Education*, 4, 513-528.

Prosser, M., & Trigwell, K. (1999). *Understanding Learning and Teaching: The Experience in Higher Education*. Philadelphia, PA: Society for Research into Higher Education & Open University Press.

Prosser, M., Trigwell, K., & Taylor, P.T. (1994). A phenomenographic study of academics' conceptions of science learning and teaching. *Learning and Instruction*, 4, 217-231.

Prosser, M., Walker, P. & Millar, R. (1994). Differences in students' perceptions of learning physics. *Research and Development in Higher Education*, 17.

Ramsden, P. (1979). Student learning and perceptions of the academic environment. *Higher Education*, 8, 411-428.

Ramsden, P. (1987). Improving teaching and learning in higher education: The case for a relational perspective. *Studies in Higher Education*, 12, 275-286.

Ramsden, P. (1988). Studying learning: Improving teaching. In Ramsden, P., (Ed.), *Improving Learning. New Perspectives* (pp. 13-31). London: Kogan Page.

Ramsden, P. (1991). A performance indicator of teaching quality in higher education: The Course Experience Questionnaire. *Studies in Higher Education*, 16, 129-150.

Ramsden, P. (1992). *Learning to Teach in Higher Education*. London: Routledge.

Ramsden, P. (1997). *Using Aims and Objectives*. Research Working Paper 89.4, University of Melbourne.

Ramsden, P., Prosser, M., Trigwell, K. & Martin, E. (1997). Perceptions of academic leadership and the effectiveness of university teaching. Paper presented at the *Annual Conference of the Australian Association for Research in Education*, Brisbane, Australia.

Richards, M., & Pelley, L. (1994). The 10 most valuable components of an information systems education. *Information and Management*, 27, 59-68.

Richards, T., Yellen, R., Kappelman, L., & Guymes, S. (1998). Information systems manager's perceptions of IS job skills. *Journal of Computer Information Systems*, 38(3), 53-57.

Roth, M.K., & Ducloss, L.K. (1995). Meeting entry-level job requirements: The impact on the IS component of business undergraduate curricula. *Journal of Computer Information Systems*, 35(4), 50-54.

Schuldt, B.A. (1991). Real-world versus "Simulated" projects in database instruction. *Journal of Education for Business*, 61(1), 35-39.

Staehr, L. (1999). Teaching ethics to computing students. *Proceedings of the First AICE International Conference*, Ed. Simpson, C. R., Melbourne, Australia, July 14-16, pp. 347-355. Full paper available online: http://www.aice.swin.edu.au/events/AICEC99/webabstracts.html#STAEHR.

Staehr, L., & Cope, C. J. (1999). A new model for teaching information systems? *Proceedings of the 3rd World Multiconference on Systemics, Cybernetics and Informatics and the 5th International Conference on Information Systems Analysis and Synthesis,* Callaos, N. et al., (Eds.). Vol. 1, Orlando, FL., USA, July 31-August 4, pp. 357-362.

Sullivan, S. L. (1993). A software project management course role-play team-project approach emphasizing written and oral communication skills. *SIGCSE Bulletin,* 25(1), 283-287.

Svensson, L. (1984). Skill in learning. In Marton, F., Hounsell, D. and Entwistle, N. J. (Eds.), *The Experience of Learning,* 56-70. Edinburgh: Scottish Academic Press.

Thomson, N.S. (1994). Using TQM principles to teach current topics in information systems. *Journal of Information Systems Education,* 6(2), 65-72.

Trauth, E.M., Farwell, D.W., & Lee, D. (1993). The IS expectation gap: Industry expectations versus academic preparation. *MIS Quarterly,* 17(3), 293-307.

Trigwell, K., & Prosser, M. (1991). Improving the quality of student learning: The influence of learning context and student approaches to learning on learning outcomes. *Higher Education,* 22, 251-266.

Trigwell, K., & Prosser, M. (1996). Changing approaches to teaching: A relational perspective. *Studies in Higher Education,* 21, 275-284.

Wagner, W. P. (1997). Teaching information systems concepts using World Wide Web-based content. *Journal of Computer Information Systems,* 38(2), 76-81.

Yaverbaum, G.J. (1987). An evaluation of a realistic approach to MIS. *SIGCSE Bulletin,* 19(1), 36-39.

Chapter V

Computer-Supported Learning of Information Systems: Matching Pedagogy With Technology

Raquel Benbunan-Fich and Leigh Stelzer
Seton Hall University, USA

ABSTRACT

New information technologies (IT) can enhance management information systems (MIS) education by improving the quality of the learning experience. This chapter proposes a tri-dimensional conceptual model based on the pedagogical assumptions of the course, the time dimension of the communication between students and professors, and the geographical location of learners and instructors. The implications of the model are reviewed in terms of their potential to contribute to teaching MIS courses and doing research in computer-supported MIS education.

INTRODUCTION

Educational applications of Information Technology (IT) have received increased attention in the academic literature (e.g., Alavi, Wheeler, & Valacich, 1995; Alavi, Yoo, & Vogel, 1997; Hiltz, 1994; Webster & Hackley, 1997) and in the business and professional press (e.g., Hibbard, 1998). Information systems classes are the source of many groundbreaking studies, though IT is being applied

Copyright © 2002, Idea Group Publishing.

to teach virtually every academic subject. Most studies focus on how IT can enhance or substitute for traditional classroom activities. However, with very few exceptions such as Leidner and Jarvenpaa (1995), little attention has been paid to the development of systematic models to guide successful applications of IT in educational settings. There is a need for a systematic appraisal of the possible learning gains when pedagogy is leveraged with technology.

Computer-supported learning (CSL) environments promise to enhance traditional teaching models and permit the exploration of new teaching/learning paradigms that were not possible before. CSL is the use of a computer connected to the Internet or an intranet (restricted or in-house network) to provide instruction to students on the network. CSL involves taking advantage of the computer's graphic and communication capabilities, including presentations, simulations, chat-rooms, email and testing software. CSL can be used in a traditional university classroom setting, a corporate classroom or at the "home" desk of a student or worker learner.

A learning revolution based on the connectivity of the Internet and the potential of the World Wide Web is underway. Information technology and management information systems (MIS) courses, as well as courses in many other subject areas, are being transformed by the prospect of harnessing IT inside and outside the classroom. The aim of this chapter is to review theoretical and empirical research on CSL, examine the implications for MIS education and present an alignment framework to focus instructional and research thinking. The three driving forces of this framework are the pedagogical model driving the course, the time dimension of communication between students and professors, and the geographical location of course participants. These three factors will define a new framework to suggest and to analyze pedagogically effective situations in MIS education.

BACKGROUND

Computer-supported learning has received increased attention from educators and researchers as a result of the convergence of technology and economics (Chatterji, 2000). In technology, rapid advances in telecommunications are linking not only individual students with their peers and instructors, but also entire schools with their counterparts across the globe. These changes are fueled by the increased affordability of personal computers and the explosive growth of computer-mediated communications that are responsible for the larger number of computers in households, campuses and offices.

At the economic level, colleges and universities are facing declining resources and are looking for ways to reduce costs or to expand their markets (Alavi, Yoo, & Vogel, 1997). For them, computer-supported learning options represent an opportunity and a threat (Twigg, 2000). On the one hand, they can potentially attract more students or increase class sizes by using the technology. But on the other hand, they are competing with their own traditional offerings as well as other educational institutions and even commercial learning companies. Increased competitive pressures along with the availability of technology are producing a shift from classroom-based education to computer-supported learning.

In addition to these technological and economic trends, the lifelong learning and continuous education movements have moved responsibility for learning from the professor/instructor to the student/worker (Farrington, 1999). In this context, the possibility of achieving more efficient teaching and better learning with CSL is very attractive. Thus the stage is set for computer-supported learning.

Teaching/Learning Models

There are two opposite but complementary models of the teaching/learning process. Teaching can be one-way "objectivist" transmission of knowledge from the professor to the students in which each student learns individually (Bouton & Garth, 1983). Alternatively, teaching can be a communal experience where knowledge is created through constructive dialogue and group discussion, in which the professor *facilitates* the process. These different approaches make different demands on technology.

The *instructive* model of learning (objectivist) assumes that there is a single objective reality and the goal of learning is to understand that reality. In this instructor-centered model, the professor is someone who has attained a high degree of understanding and knowledge in the field and the student is a passive recipient of instruction.

By contrast, the *constructive* perspective assumes that knowledge is created or constructed by every learner interacting with others (Leidner & Jarvenpaa, 1995). Learning is based upon a model that views the student as an active participant in individual or group activities designed to construct or create knowledge. In fact, collaborative (or group) learning is one of the most important implementations of the constructive approach. Collaborative learning refers to instructional methods that encourage students to work together on academic tasks and engage in problem solving that promotes learning (Harasim, 1990).

In collaborative activities, learning is a social process that takes place in communication with others, not inside the mind of the individual student. It requires dialogue and the development of interpersonal skills. Learning occurs because of the collaboration of a group of students, rather than because of their isolated efforts as individuals. Knowledge is viewed as a social construct, generated by peer interaction, discussion, reaction, evaluation and cooperation. Therefore, the role of the professor changes from transferring knowledge to students to guiding students' construction of their own knowledge.

Collaborative learning is associated with a number of benefits. According to Webb (1982), participating in a group promotes positive socio-emotional effects and cognitive growth that contributes to learning. At the emotional level, groups create a climate favorable to learning by increasing motivation and reducing anxiety. At a cognitive level, working in a group develops higher order mental processes. When a group solves problems, mental models are challenged, extended and refined (Shuell, 1986) and higher order cognitive skills are developed. The group learns by applying learned concepts to novel situations, thinking critically and synthesizing diverse materials (Hiltz, 1994). Teamwork can produce the externalization of the thought processes, the comparison of alternative perspectives, social facilitation, better learning, high self-esteem and more positive attitudes toward the learning experience (Salomon & Globerson, 1989).

There is, however, a downside potential. As in any other group endeavor, collaborative learning may be affected by social-loafing, coordination problems and other process losses (Nunamaker et al., 1991). Rather than pooling their mental efforts, some team members may actually show reduced expenditure of mental effort, loafing behavior and even effort avoidance in ways that undermine learning (Salomon & Globerson, 1989).

Both the instructive and constructive models suggest that teaching is at its core a communication process. The instructive paradigm conceives teaching as a one-way transmission of information while the constructive perspective presents teaching as an interactive dialogue between learners and instructors or among learners. Technology can be employed for these two different purposes: (1) to transmit content (to deliver instruction), and (2) to provide a platform for communication between professors and students, and among students. As a one-way transmission tool, IT can complement or supplant instructor lectures and printed materials. As a multi-directional communication platform, IT can extend faculty availability beyond class times and office hours and establish links among classmates (Benbunan-Fich, 1999).

Teaching MIS Courses

There are two main challenges associated with teaching MIS courses. The first is the need to adapt the content to the rapid advances of the technology. Each IT environment whether it is mainframe, client/server or Web-based poses new and different issues for the MIS discipline. Since MIS is primarily a pragmatic field, the second challenge is the need to bridge the gap between the theory and practice of information systems.

Collaborative and constructive pedagogical techniques allow MIS instructors to add flexibility to their courses and adapt them to new technologies. At the same time, these pedagogical techniques permit instructors to illustrate concepts from a pragmatic standpoint. The case study method, widely used in MIS courses, is an example of a collaborative pedagogical technique. Case studies present real or hypothetical situations that demand group discussion and the use of concepts to develop recommendations and generate preferred solutions. Through this activity, students process instructional inputs, assimilate course materials and bridge the gap between the theory and practice of information systems.

MIS courses are also appropriate settings for other collaborative learning activities such as debates, group projects, simulation and role-playing exercises. When students participate in these activities, they can experience active learning, teamwork and higher cognitive processes. The combination of these three factors in appropriate educational settings is more effective and engaging than the usual one-way transmission of concepts (Harasim, 1990; Hiltz, 1994; Johnstone, 1991).

Due to the continuing developments in technology, MIS instructors should use flexible learning methods based on interaction and collaboration combining traditional classroom experiences with online interaction. In this context, recent advances in computer-mediated communications are removing time and space constraints and offering new possibilities for adding flexibility to MIS courses (Farrington, 1999).

Time and Place Classification

Communication technologies can provide time-independent communication and place-independent communication. By overcoming time constraints, these systems facilitate self-pacing and self-directed learning, and they can improve in-depth reflection, comment formulation and topic development (Harasim, 1990). By lifting geographical constraints, remote students can be members of a team. All students have expanded access to peers, experts and sources of information regardless of their geographical location.

Computer networks support two modes of instructional delivery, real-time and delayed time instruction. **Synchronous** interaction takes place when instructors broadcast live lectures over the Web (Web-cast their lectures) and take student questions in real-time. In other words, students and instructors work together at the same time. By contrast, **asynchronous** interaction occurs when learners and professors work at different times. Since they do not need to work simultaneously, their interaction is independent of time. In these asynchronous electronic classrooms, students and professors rarely meet face-to-face. Their interaction is completely mediated by computer networks. An implementation of this approach is called **Asynchronous Learning Networks (ALN)** (Hiltz & Wellman, 1997). The ALN is tailored to mediate learning processes and to support "anytime/anywhere" interaction by providing a combination of database, e-mail and conferencing capabilities.

Internet or Web-based delivery usually occurs through a comprehensive web site that combines a repository of information and communication-support tools. As a source of content, the Web provides access to traditional archives and to the worldwide information on millions of Web sites (Kuechler, 1999). In addition to replacing conventional instruction materials, the Internet can complement and expand traditional communication channels between students and professors and among students (Benbunan-Fich, 1999).

However, using computer networks to enhance the instruction delivery has the advantages of easy updating of instructional materials and elimination of costly printed materials. Intranet and Internet-based instruction also provide the immediate distribution that is not possible with other media. In addition to instant flexibility, network technology unleashes the power of collaboration and allows students or workers to learn through formal and informal interaction with their peers and instructors (Hibbard, 1998).

Interaction along the time and place dimensions can be described in terms of the fourfold typology proposed by Johansen (1992). Interaction can occur at the same time (synchronous) or at different times (asynchronous). Members can meet in the same place (proximate) or in different places (distributed). Figure 1 presents Johansen's typology.

Same time/same place refers to teams that meet in the same room or labs in which each student has access to a personal computer. **Same time/ different place** situations occur when team members are located in (at least two) different places and communicate via computers using chat rooms or desktop-video conferences. The "same time" cells lend themselves mostly to

Figure 1: Time and Place Classification

	Time	
	Same	Different
Same Place	Synchronous/Proximate	Anytime/Same (virtual) Place
Different	Synchronous/Dispersed	Asynchronous/Dispersed

Source: Adapted from Johansen (1992)

regional teaching, due to the difficulty to coordinate simultaneous participation among people in different time zones.

Different time/same place category normally refers to people who work in shifts but share a common meeting room (project room) where they leave messages for each other and share materials. In this category, Chizmar and Williams (1996) expand the notion of place to include dedicated virtual workplaces restricted to a particular class or group of students. An example is a course homepage on the web used as a shared virtual space to store class materials.

Finally, **different time/different place** refers to asynchronous-dispersed teams that rarely meet face-to-face and transact all their work through an asynchronous communication system. This category seems to be attracting most of the attention at the institutional level because it is the one that enables universities to globalize their offerings and expand their markets.

ALIGNMENT FRAMEWORK

To be effective, educational applications of IT must be aligned with the pedagogical model driving the course. The constructive paradigm makes different demands on technological support in education than does the instructive model. Pedagogy emphasizing the "one-way" transmission of concepts suggests improvements in the efficiency and effectiveness of lectures and class demonstrations. Constructive learning, by contrast, requires technological platforms that support interaction and communication among students (Benbunan-Fich, forthcoming).

In addition, the temporal and spatial characteristics of the communication also determine the appropriate technology. The temporal dimension is

Figure 2: Alignment Framework

Pedagogy	Location	Mode	
		Synchronous	*Asynchronous*
Instructive	*Proximate*		
	Dispersed		
Constructive	*Proximate*		
	Dispersed		

divided into synchronous (same time) and asynchronous (different time) depending on whether the communication is in real-time or in delayed-time. The spatial dimension is divided into proximate (same place) and distributed (different place) depending upon the location of course participants. The combination of the pedagogical model with the time and place classification produces a matrix that helps to categorize different educational applications of IT. Figure 2 presents the model.

IT Applications for the Instructive Approach

The centerpiece of the instructive approach is the lecture. Different IT applications that can increase the efficiency of this transmission include presentation technology (e.g., PowerPoint), tutorial software and computer-assisted instruction where each student works directly with the computer to learn and practice the material (Chambers & Sprecher, 1980). IT also allows participants in remote locations (students, professors or guest speakers) to take part in the same lecture carried out in real-time and broadcast to everyone via computers enhanced with video, audio and data links. In this setting, the use of chat rooms enable participants to join class discussions in real-time (Alavi, Wheeler, & Valacich, 1995; Webster & Hackley, 1997).

The traditional lecture can be detached from the same time/same place notion enabling the professor to extend class time. Real-time or previously videotaped lectures can be distributed via computer to the students who can access streaming media at their own convenience. The students can get these "e-lectures" anytime/anywhere. An ALN enables participants to have class discussions, to receive assignments and to submit papers.

Another virtual extension of lectures is a course Web page containing lecture notes and class materials (Chizmar & Williams, 1996). The course Web page can be a comprehensive repository of course-related materials including: the syllabus, assignments, readings, class notes, study guides, selected papers and general announcements that are updated regularly (Kuechler, 1999). Students can visit the homepage to download class materials or assignments anytime.

The upper part of Figure 3 classifies the different IT applications to enhance the instructive method. The synchronous/proximate category features presentation technology and tutorial software. The synchronous/dispersed cell corresponds to networked classrooms via video or data links, chat rooms and desktop videoconferences. The asynchronous/proximate category refers to the use of a course Web site as a repository of class materials. Finally, the asynchronous/dispersed category contains different modes of anytime/anywhere lecture delivery.

IT Applications for the Constructive Pedagogy

In the constructive context, IT applications range from the use of Group Support Systems (GSS), or desktop video conferences for groups meeting in real-time (Alavi, Yoo, & Vogel, 1997), to e-mail and ALN to collaborate with other students independent of time and place (Benbunan-Fich & Hiltz, 1999). These communication technologies introduce powerful environments to enhance social and intellectual connectivities (Harasim, 1990) by structuring information exchange, supporting peer and faculty-student interaction, and overcoming time and space barriers.

Collaborative learning assignments are appropriate to every cell of the time/place classification using synchronous Group Support Systems (GSS), chat rooms and/or ALN, depending on the temporal and spatial dimensions of the communication. Groups meeting face-to-face to discuss case studies can use a synchronous GSS in a decision room. Groups with remote participants meeting in real-time can use chat rooms or desktop-video conferences for discussions. And groups meeting asynchronously can use e-mail or an ALN to support their electronic discussions.

When used for group problem solving, synchronous and asynchronous computer-mediated communication systems have several advantages. First, they may enhance the collaborative learning experience by moderating social pressures associated with face-to-face participation such as: turn taking (fragmenting the available speaking time), dominating or monopolizing discussion, and fear of reprisals and cognitive inertia (Nunamaker et al., 1991). Second, everyone is perceived similarly regardless of physical handicaps, regional and national accents, usual assertiveness in face-to-face discussions and other characteristics that tend to advantage/disadvantage students in the usual classroom setting (Johnstone, 1991). Third, the system organizes the contributions and maintains a written transcript of the interaction that can be revised, stored and subsequently retrieved (Harasim, 1990). Fourth, the need to interact using written expression can enhance meta-cognitive skills such as reflection and revision (Harasim, 1990).

The disadvantages of computer-mediated group communication outside the classroom are a function of the temporal and spatial dimensions of the communication. For example, in synchronous discussions (GSS), participants may be affected by cognitive inertia, a tendency to think along the same lines, or production blocking - an inability to produce meaningful contributions (Nunamaker, et al. 1991). In asynchronous discussions (ALN), students may engage sporadically or experience anxiety from delays and different participation rates (Benbunan-Fich and Hiltz, 1999).

The lower portion of Figure 3 presents the relevant IT applications for the constructive approach. Synchronous/proximate situations can employ a GSS in a decision room to support constructive and collaborative learning activities. The synchronous/dispersed cell can use networked classrooms, chat rooms or desktop videoconferences to complete collaborative assignments that would be much more onerous without these technologies. IT applications in the asynchronous/dispersed category range from e-mail to computer conferences and ALNs to support constructive learning anytime/anywhere. The synchronous/dispersed category features a novel application combining case studies with simulations by using interactive web-based cases with hypertext environments. These Web cases are located in the same virtual place and can be visited by students at different times. For example, HyperCase® is an interactive web-based case where students explore MIS problems of a fictional company (Maple Ridge Engineering). MRE HyperCase can be found at: http://www.thekendalls.org. Playing the role of a systems analyst, students interview workers, visit offices and analyze documents (Kendall et al., 1996).

New developments such as the convergence of voice, data and video into a common digital infrastructure and the increased connectivity between wired and wireless networking will offer even more possibilities for each cell of the framework. Educational uses of computer networks require considerable investments of time

Figure 3: IT Applications in the Alignment Framework

Pedagogy	Location	Mode	
		Synchronous	*Asynchronous*
Instructive	*Proximate*	**Presentation technology and tutorial software** (Leidner & Jarvenpaa, 1995)	**Course Web page** (repository of lecture notes and class materials) (Kuechler, 1999)
	Dispersed	**Real-time linked classrooms** (Alavi, Yoo and Vogel, 1997)	**Virtual classroom** (Hiltz, 1994) **and Web-telecourses** (LaRose et al., 1998)
Constructive	*Proximate*	**Synchronous GSS in decision rooms** (Alavi, 1994)	**Web-based Interactive case studies** (Kendall et al., 1996)
	Dispersed	**Remote GSS, chat rooms and desktop video-conferences** (Alavi et al., 1995)	**Asynchronous Learning Networks (ALNs)** (Benbunan-Fich & Hiltz, 1999)

and effort, not to mention the monetary costs of the hardware/software infrastructure and the technical support. Although a cost-benefit analysis of each application is beyond the scope of this chapter, instructors must carefully plan their innovations to avoid frustration and the danger that technical aspects take center stage (Schweizer, 1999).

IMPLICATIONS

The alignment framework highlights the appropriate match of technological possibilities with pedagogical models. In the context of MIS classes that typically feature both instructive and constructive pedagogies, the framework defines a selection of IT applications that improve learning effectiveness and increase teaching efficiency. The framework can also facilitate review and classification of the empirical research on MIS education. From the many studies conducted in computer-supported learning, a sample of research conducted in Computer and Information Systems courses has been selected to illustrate each cell of the framework. For each study, we will briefly describe the methodology and highlight the results regarding actual learning and perceived learning.

Research in the Instructive Matrix

Research in this domain usually compares novel lecturing approaches with traditional ones. For example, Alavi, Yoo and Vogel (1997) investigated networked classrooms at two universities (same time/different place). Results indicated no significant differences in mastery of the material between face-to-face and videoconference lectures, but students were less satisfied with the videoconference lectures.

New modalities of asynchronous lecturing have also been the subject of empirical projects. A longitudinal study compared traditional classrooms with totally distant classrooms using a combination of videotaped lectures and ALN-supported communication ('Virtual Classroom™' in Hiltz, 1994). In terms of final grades, there were no significant differences between traditional (face-to-face, not computer-supported) courses and VC-supported classes in mastery of the material. Subjectively, however, most students reported that VC was overall a better way of learning than traditional classes.

LaRose, Gregg and Eastin (1998) studied a web telecourse in which students listened to pre-recorded audio classroom interactions while viewing a detailed course outline and illustrative web sites. Findings showed that telecourse students' test scores and perceptions were equal to those in a comparison traditional classroom. Thus it appears that lecturing in real-time to networked locations or delivering asynchronous lectures by videotape or the Web do not affect students' mastery of the material.

Research in the Constructive Matrix

Several studies compare the effectiveness of solving MIS case studies in different technologically supported situations. Alavi (1994) compared groups using a GSS to face-to-face unsupported groups of MBA students. GSS groups expressed higher levels of perceived skill development, self-reported learning and evaluation of their classroom experience. Although there were no significant differences in midterm scores, final test grades for students in the GSS groups were significantly higher than grades of students without CSL.

Alavi, Wheeler and Valacich (1995) conducted a longitudinal field study to compare the effectiveness of three collaborative learning environments: computer-mediated proximate groups, computer-mediated non-proximate groups (using desktop video-conference) and face-to-face unsupported groups. They found that the three situations were equally effective as measured by students' knowledge acquisition and satisfaction with process and outcomes. However, the students in the non-proximate groups using desktop video-conferencing showed higher critical thinking skills and greater commitment to their groups than students in the other two conditions.

In the asynchronous domain of this matrix, Benbunan-Fich and Hiltz (1999) sought to determine the separate and joint effects of communication media and teamwork. They compared groups and individuals solving a computer ethics case with and without an ALN. The ALN appeared to enhance the quantity and quality of the solutions to an ethical case scenario. The combination of teamwork with ALN-support increased the students' perception of learning, but not final exam scores. ALN-supported groups were less satisfied than their real-time counterparts.

Ocker and Yaverbaum (1999) reported results of a repeated-measures experimental study of MIS case-study problem solving that compared student groups using traditional face-to-face collaboration with groups using asynchronous collaboration through conferencing software. Results for the asynchronous collaboration equaled those for face-to-face collabo-ration for learning, quality of solution, solution content and satisfaction with the solution quality. However, students in asynchronous groups were less satisfied with the group interaction process and the quality of group discussions.

It is still early days for research that directly compares synchronous and asynchronous collaborative learning approaches. Yet, preliminary studies suggest that communication technologies enhance collaborative learning approaches and, in particular, the case-solving process. Asynchronous groups, however, tend to be less satisfied than their face-to-face or real-time counterparts. These consistencies persist despite the use of different communication technologies. The alignment model proposed in this chapter enhances the prospects for future studies. It will facilitate the design of comprehensive studies to test different options for supporting MIS education with the new technologies and will provide a unified framework for comparing results and advancing knowledge in the field.

CONCLUSIONS

There are many possible applications of IT in MIS education, but there is a lack of theoretical frameworks to classify these applications. As a contribution in this direction, this chapter has integrated three relevant dimensions in a conceptual model that illustrates different alternatives. IT support for MIS education should be driven by the pedagogical model used in the course, the temporal dimension of the communication and the spatial location of the participants. A typical MIS course features elements of both the instructive and constructive models. Concepts are transmitted mainly through lectures, and collaborative assignments are used to reinforce learning.

The interest in anytime/anywhere instruction conjoins with the development of technologies that are overcoming temporal and spatial barriers. Instructors now can identify instructional technologies specific to their pedagogic model and their temporal and spatial requirements. Effective applications of IT in education are those aligned with the teaching/learning model driving the course. At the pedagogical level, the model can be used as a road map to select the most effective IT-supported learning activities in order to enhance MIS education. In addition, the model also provides a context for theory and research in computer-supported learning, a categorization for previous findings and a guide for empirical research studies.

REFERENCES

Alavi, M. (1994). Computer-mediated collaborative learning: An empirical evaluation. *MIS Quarterly*, June, 150-174.

Alavi, M., Wheeler, B., & Valacich, J. (1995). Using IT to reengineer business education: An exploratory investigation to collaborative tele-learning. *MIS Quarterly*, September, 294-312.

Alavi, M., Yoo, Y., & Vogel, D. (1997). Using information technology to add value to management education. *Academy of Management Journal*, 40(6), 1310-1333.

Benbunan-Fich, R. (forthcoming). A conceptual model to improve education and training with information technology. *Communications of the ACM*.

Benbunan-Fich, R. (1999). Leveraging management education with information technology. In Elsass, P., (Ed.). *Proceedings of the Eastern Academy of Management Annual Meeting*, May 10-13, Philadelphia, PA. 223-226.

Benbunan-Fich, R., & Hiltz, S. R. (1999). Impacts of Asynchronous Learning Networks on individual and group problem solving: A field experiment. *Group Decision and Negotiation*, 8(September), 409-426.

Bouton, C., & Garth, R. Y. (1983). *Learning in Groups*. San Francisco, CA: Jossey-Bass, Inc.

Chambers J. A., & Sprecher, J. W. (1980). Computer assisted instruction: Current trends and critical issues. *Communications of the ACM*, 23, 332-342.

Chatterji, D. (2000). Strategic impact of the information age. *Research Technology Management*, 43(1), 35-37.

Chizmar, J., & Williams, D. (1996). Altering time and space through network technologies to enhance learning. *Cause/Effect*, 19(3), 14-21.

Farrington, G.C. (1999). The new technologies and the future of residential undergraduate education. In Katz, R. et al., *Dancing with the Devil: Information Technology and the New Competition in Higher Education*. San Francisco, CA: Jossey-Bass Publishers.

Harasim, L. (1990). *Online Education: Perspectives on a New Medium*. New York: Praeger/Greenwood.

Hibbard, J. (1998). The learning revolution. *Information Week*, March 9, 44-60.

Hiltz, S.R. (1994). *The Virtual Classroom: Learning without Limits via Computer Networks*. Norwood, NJ: Ablex Publishing Corporation.

Hiltz, S. R., & Wellman, B. (1997). Asynchronous Learning Networks as a virtual classroom. *Communications of the ACM*, 40(9), 44-49.

Johansen, R. (1992). An introduction to computer-augmented teamwork. In Bostrom, R. P., Watson, R. T., and Kinney, S. T. (Eds.), *Computer Augmented Teamwork: A Guided Tour*. New York, NY: Van Nostrand Reinhold.

Johnstone, S. (1991). Research on telecommunicated learning: Past, present and future. *The Annals of the American Academy of Political and Social Science*. 514(March), 49-57.

Kendall, J. E., Kendall, K. E., Baskerville, R., & Barnes, R. (1996). An empirical comparison of a hypertext-based system analysis case with conventional cases and role-playing. *The Data Base for Advances in Information Systems*, 27(1), 58-77.

Kuechler, M. (1999). Using the Web in the classroom. *Social Science Computer Review*, Summer, 144-161.

LaRose, R., Gregg, J., & Eastin, M. (1998). Audiographic telecourses for the Web: An experiment. *Journal of Computer-Mediated Communication*, 4(2). [On-line] Available: http://www.ascusc.org/jcmc/vol4/issue2/larose.html.

Leidner, D. E., & Jarvenpaa, S. (1995). The use of information technology to enhance management school education: A theoretical view. *MIS Quarterly*, September, 265-291.

Nunamaker, J., Dennis, A., Valacich, J., Vogel, D., & George, J. (1991). Electronic meeting systems to support group work. *Communications of the ACM*, 34(7), 41-61.

Ocker, R. J., & Yaverbaum, G. J. (1999). Asynchronous computer-mediated communication vs. face-to-face collaboration: Results on student learning, quality and satisfaction. *Group Decision and Negotiation*, 8(September), 427-440.

Salomon, G., & Globerson, T. (1989). When teams do not function the way the ought to. *Journal of Educational Research*, 13(1), 89-100.

Schweizer, H. (1999). *Designing and Teaching an On-line Course*. Boston, MA: Allyn and Bacon.

Shuell, T. (1986). Cognitive conceptions of learning. *Review of Educational Research*, 56(4), 411-436.

Twigg, C. A. (2000). Institutional readiness criteria. *EDUCAUSE Review*, 35(2), 42-51.

Urdan, T. A., & Weggen C. A. (2000). *Corporate E-Learning: Exploring a New Frontier*. WR Hambrect & Co.

Webb, N. M. (1982). Student interaction and learning in small groups. *Review of Educational Research*, 52(3), 421-445.

Webster, J., & Hackley, P. (1997). Teaching effectiveness in technology-mediated distance learning. *Academy of Management Journal*, 40(6), 1282-1309.

Chapter VI

Problem-Based Learning in Information Systems Analysis and Design

John Bentley and Geoff Sandy
Victoria University, Australia

Glenn Lowry
United Arab Emirates University

ABSTRACT

The critical question challenging information systems educators in the new millennium is how university information systems courses can add enough value to students that they will choose to study in higher education for a full degree rather than opt for a one-year certification course leading to similar economic and status outcomes in the short term. This chapter assesses the feasibility and desirability of achieving a better match between delivery of information systems education and the professional workplace through Problem-Based Learning (PBL). A brief introduction to cognitive and learning principles is followed by a discussion of PBL and its potential to help to achieve a better fit between student aspirations and employer requirements. The chapter concludes with an illustration of the use of PBL in a systems analysis and design course.

Copyright © 2002, Idea Group Publishing.

INTRODUCTION

As the 21st century gains momentum, university information systems departments face increasing competition from IT industry vendors in the education of new professional entrants. Several major vendors have developed "certification" schemes in the past five years, largely in response to a shortage of people trained in the application and use of their products (NOIE, 1998, 1999; Orr & von Hellens, 2000). They have concluded that the growth of their firms and the sales of their products are severely limited by a worldwide shortage of properly trained technical staff.

At the same time, vendors such as Microsoft, Oracle and SAP have enthusiastically worked to initiate and support "strategic partnerships" with university information systems and computing departments. The vendors would rather *not* be in the university education business. They acknowledge that universities do that better than they do. They would prefer to concentrate their efforts and resources in their core (and more lucrative) businesses.

Universities may have begun to lose their short-lived and always shaky monopoly on professional entry into information systems careers. A growing number of aspiring young IT professionals are selecting certification by software vendors over university study as a means of preparation and career entry. After one year of focused technical study, followed by an inexpensive, independent examination, many newly certified vendor-trained "solution developers," "systems administrators" and "Web site developers" command salaries and working conditions equal to or, often, superior to new information systems graduates.

Prins Ralston, a recent president of the Australian Computer Society, commented on this issue in the media (Ralston, 1999), stating that:

> "One of the things that has come out of the IT skills crisis is the need for our universities to produce the best students possible, young people equipped with the knowledge, technical and social skills to move straight into vacant positions within industry and commerce."

Turner and Lowry (1999) address the issue of reconciling the career aspirations of business information systems graduates with the intellectual skills and personal attributes desired of them by employers. Finding some conflict between the interests of these stakeholders, the authors went on to identify factors that motivate intending information systems graduates seeking initial employment (Turner and Lowry, 2000, 2001). The crucial issue for information systems educators is to achieve a "correct" balance between these sometimes conflicting interests.

Students commonly represent themselves as primarily concerned about securing a job following graduation and with subsequent career advancement.

They look to tertiary educators to teach them technical knowledge and skills which will optimize their chances of achieving this goal. Many tend not to appreciate the importance placed by employers on the ability for higher order thinking and interpersonal skills such as oral and written communication or the ability to work in a team. On the other hand, employers often lament the fact that business information systems graduates lack these "soft" abilities.

Educators may well question whether the expectations of new graduates of employers are realistic. Graduates, who three or four years previously were secondary school leavers, will lack "polish" or the political sensitivity that only come with maturity and organizational experience. We may also question whether some of the personal attributes of graduates desired by employers, like curiosity or self-discipline, are capable of being taught and learned. Alternatively, how can educators convince students to embrace *education*, with its emphasis on development of self-evaluative, self-educative habits, rather than *training* which is seen as a means to a "meal ticket." These issues have implications for tertiary educators regarding curriculum content, the learning environment and especially the method of delivery that is adopted.

In this chapter, we explore some of these important curriculum issues and consider whether there is a better way to educate information systems students to ensure that they become the "compleat graduate." We believe that a *student-centered* learning approach such as Problem Based Learning (PBL) may be a better way. It is an approach that has been used successfully in other disciplines such as medicine and law (Boud and Feletti, 1991; Savery and Duffy, 1995). To our knowledge, the PBL method has not been meaningfully adopted for information systems education.

In the sections that follow, we will present a model of the "Compleat Business Information Systems Graduate," followed by a discussion of the nature of PBL and how it differs from other approaches. The chapter concludes with an illustration of the use of PBL in a systems analysis and design course.

A Model of the Compleat Information Systems Graduate

Curricula for undergraduate computing programmes have been developed over many years by the relevant professional associations like IEEE, ACM and DPMA. These overseas associations have significantly influenced development of curricula in Australia. Recently, the Australian Computer Society (ACS) published a document identifying the core body of knowledge for an information technology professional (Underwood, 1997). Turner and Lowry (1999) in their Australian study found that:

- Employers expect new graduates to be immediately productive.

- Students are aware of the need to possess certain knowledge and skills to obtain employment and for career advancement.
- Students do appreciate the need for interpersonal skills and personal attributes that are highly valued by employers.

To guide ourselves and to assist tertiary educators in determining content of the curriculum, and in choosing the best delivery mechanism, a simple working model for developing the "compleat information systems graduate" is shown in Figure 1 below.

Figure 1 suggests a developmental process in which personal attributes, which influence intellectual abilities and skills, are applied to acquisition of knowledge to enable the development of higher cognitive activities. Figure 1 is consistent with Bloom's *Taxonomy of Educational Objectives: The Classification of Educational Goals,* one of the enduring foundations of learning theory.

Bloom (1956) identified six cognitive domains and described their associated cognitive activities. The cognitive domains are arranged in a hierarchy of increasing sophistication. Knowledge is the lowest cognitive domain, but higher order thinking is based upon knowledge of some of the "realities." Table 1 summarizes Bloom's *Taxonomy of Educational Objectives.*

Bloom emphasized the importance of intellectual abilities and skills. When these are applied to knowledge, it facilitates higher order cognitive activities. The educative process is about inducing change in students, that is, change in their thinking. At the end of the educational process, students must be able to *do something* with their knowledge. They must be able to apply knowledge to new situations and problems. This requires certain generic intellectual abilities and skills.

The model in Figure 1 also identifies as important something that is referred to as "personal attributes." Mulder (1998) indicated that employers listed attributes like curiosity, risk taking, personal discipline and persistence high among the desirable personal "skills" of an information technology/information systems graduate. These attributes can influence in important ways the successful application of

Figure 1: A Model of the Compleat Information Systems Graduate

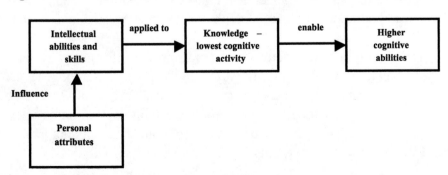

Table 1: A Summary of Bloom's Taxonomy of Educational Objectives

Cognitive Domain	Cognitive Activity
Evaluation	Judgments about the value of material and methods for given purposes. Quantitative and qualitative judgments about the extent to which material and methods satisfy criteria. Use of a standard of appraisal.
Synthesis	The putting together of elements and parts so as to form a whole. Combining them in such a way as to constitute a pattern or structure not clearly there before.
Analysis	The breakdown of a communication into its constituent elements or parts such that the relative hierarchy of ideas is made clear and/or the relations between the ideas expressed are made explicit.
Application	The use of abstractions in particular and concrete situations. The abstractions may be in the form of general ideas, rules of procedures or generalized methods. The abstractions may also be technical principles, ideas, and theories that must be remembered and applied.
Comprehension	Understanding or apprehension such that the individual knows what is being communicated and can make use of the material or idea being communicated without necessarily relating it to other material or seeing its fullest implications.
Knowledge	The recall of specifics and universals, the recall of methods and processes, or the recall of a pattern, structure or setting.

the intellectual skills and abilities to knowledge to support the higher orders of thinking. However, an important educational consideration is whether these attributes can be taught and learned. If they cannot be learned, can they be influenced or enhanced?

The knowledge and intellectual skills required of an information systems graduate are neither contentious nor difficult to identify. The same may be generally said for desired personal attributes and interpersonal skills. The critical issue and the one *likely to be contentious* is the appropriate learning environment to best prepare graduates to engage in higher order thinking. A student-centered approach such as PBL may be a better way than those currently in use.

THE NATURE OF PROBLEM-BASED LEARNING

In the view of Ryan et al. (1996), "higher education institutions are under increasing pressure to better prepare graduates for the world of work." Reform of curriculum delivery is required to achieve the "compleat business information systems graduate." Current educational delivery paradigms are characterized as "teacher-enabled learning" and "student-enabled learning." **Teacher-enabled learning** is characterized by didactic teaching, passive learning, the teacher as the "expert," with the control of learning resting with the academic. The learning of the students is directed by the academic and is often based on what the academic believes the student needs to learn. Savery and Duffy (1995) suggest that this approach can lead to students focusing on determining what knowledge they require to pass the subject rather than the instructional objectives of the program, a phenomenon well-known to educators. In contrast, **student-enabled learning** places the student into active, self-directed learning, learning by enquiry and ownership of the learning process and goals. PBL is an active learning strategy that may be suitable for better preparing information systems students for professional practice. In the problem-based approach, complex, real-world problems or cases are used to motivate students to identify and research concepts and principles they need to know in order to progress through the problems. Students work in small learning teams, bringing together collective skill at acquiring, communicating and integrating information in a process that resembles that of inquiry.

Didactic learning, that is "teaching what is known," cannot be achieved in the information systems discipline as knowledge is increasing rapidly and as adding more into an already crowded curriculum is difficult (Lowry and Doroshenko, 1996; Lowry and Morgan, 1996). Information systems professionals work in environments that are project-oriented and problem and task based. Hence, students should be able to determine "what they know," identify "what needs to be known" and how to "gain the knowledge" to solve a problem or task. Learning how to acquire and apply these skills is part of the PBL approach. The PBL approach should provide students with "life-long" learning skills suitable to the IS profession. PBL offers an approach that can lead to a better alignment between educational delivery and the information systems professional work environment.

A working group report by Ellis et al. (1998) suggests that PBL suits the information systems and computing fields as:

- IS and computing are, for the most part, problem driven.
- Life-long learning is necessary due to the rapidly and continually changing nature of the industry.

- Practitioners must constantly update their skills and competencies in order to keep abreast of new technology.
- The project group is the predominant mode of operation within the industry;
- IS and computing cross discipline boundaries.

According to Woods (1996, p.12): "PBL is about learning subject knowledge in the context of using and developing process skills." Boud and Feletti (1991, p.14) describe PBL as:

> "... a way of constructing and teaching courses using problems as the stimulus and focus for student activity. It is not simply the addition of problem-solving activities to otherwise discipline-centred curricula, but a way of conceiving of the curriculum that is centred around key problems in professional practice. Problem-based courses start with problems rather than an exposition of disciplinary knowledge."

A large body of knowledge relating to PBL, especially in the disciplines of Health and Law, is available.

Although there are a wide variety of strategies for implementing PBL models (Savery and Duffy, 1995; Woods, 1996; Ellis et al., 1998), the PBL *problem* or *case* is the primary teaching mechanism. Saarinen-Rahiika and Binkley (1998) identify three approaches to PBL from the literature: completely integrated PBL curricula, transitional curricula, and a single-course (subject) approach. There is some debate in the literature about what constitutes PBL and the strategies for delivery (Aldred et al., 1997).

A defining procedure in PBL is the presentation of the problem before any knowledge is given. The knowledge is acquired and applied back to the problem. Students must identify what knowledge needs to be gained to solve the problem, acquire the knowledge and relate it to the problem.

Table 2

Characteristics of Problem-Based Learning
Is context-based using "real-life" situations
Focuses on thinking skills (problem solving, analysis, decision making, critical thinking)
Requires integration of inter-disciplinary knowledge/skills/behaviors
Is self-directed and develops life-long learning skills
Often requires substantive interaction with clients and others who are outside the student "team"
Is shared in small groups

Savery and Duffy (1995) present a case to show that PBL is consistent with principles of instruction arising from constructivism and that PBL is an ideal approach to learning. They argue that learners must be *constructors* of their own knowledge and must be presented with active learning situations and tasks from the environment in which they will work. They are adamant that the learning objectives and resources are *not* presented with the problem as in case-based approaches, as this latter approach does not help to develop "metacognitive skills associated with problem solving or with professional life." PBL uses authentic problems from practice to motivate student learning and develop attitudes and skills required for life-long learning. Real-world problems are messy, complex and unstructured.

Authentic real-world problems in a professional context help students to take ownership of the learning experience. Ellis et al. (1998) identify a spectrum of problem definitions from the teacher constructing and fully specifying the problem, to the teacher constructing an open-ended problem, through to students constructing the problem themselves. Learning and problem solving usually takes place in small groups (five to seven students). Students acquire the process of problem solving through collaboration and the team skills of negotiation, decision-making, allocation and management of work.

The use of increasingly complex problems draws upon the higher-order thinking skills of analysis, synthesis and evaluation. After the completion of each problem, students develop a framework for use in the next problem. This concept is referred to as "scaffolding." For example, beginners in PBL need to build upon the processes of PBL rather than learning knowledge content. Learning support to assist in building these frameworks should be considered and resources provided to assist the scaffolding process (Ellis et al., 1998; Greening, 1998). The problems can be sourced or adapted from real examples. Initially the problems should be well structured to provide clues and "scaffolding" to guide students in their learning.

The role of the tutor or facilitator in PBL is ***not to inform*** but to ***coach*** or ***guide*** to assist in higher-order thinking and to challenge the thinking of students (Savery and Duffy, 1995).

Proponents, like Aldred et al. (1997), therefore claim that PBL:
- enables students to make links with other subjects in a more meaningful way;
- acknowledges the possibility of prior learning by the student;
- increases confidence;
- increases motivation;
- enables responsibility for one's own learning;
- enhances independent learning;
- motivates life-long learning;

- promotes deep rather than surface learning;
- enhances research and communication skills;
- develops analytical and problem-solving skills;
- promotes critical thought;
- enables students to learn to question and reflect on their knowledge in a way that is not possible in a more traditional mode of curriculum delivery;
- is seen to mimic the real world;
- allows students to see more relevance in their studies and for future professional practice;
- equips students with appropriate skills and attitudes for contemporary professional life;
- provides rewarding and creative experience for academics.

These benefits are valuable for the professional development of an information systems graduate, and realization of even some of them provides a strong incentive to consider the PBL approach. The PBL approach should provide a closer alignment between the undergraduate's learning environment and the real world of the practicing information systems professional. PBL appears to be an approach that can help to deliver, if not the "compleat business information systems graduate," something closer than we are currently achieving.

All these characteristics are currently used in a range of teaching and learning approaches, but PBL draws them together. Elements of each characteristic must be present for the learning approach to be characterized as PBL.

In contrast, *subject-based learning* tends to be linear and takes an approach in which students are told the theory (what to know), are directed learn it and are expected to apply it to practice (given examples to illustrate how to use it). This can be termed a "theory into practice" model. PBL however starts from practice (real-world) situations in which students first identify what they need to know, then conceptualize it by constructing/discovering the appropriate theory and finally apply it to practice in solving the problem (Woods, 1994). This is a "practice into theory into practice" cyclic model.

The assessment of student academic work developed through PBL is different from the assessment of traditionally produced work. In PBL the *process* of learning is assessed as well as the *content* learning and outcomes. There is considerable discussion in the PBL literature of forms of assessment and evaluation of PBL students. Full discussion is beyond the scope of this chapter.

As PBL focuses more on the educational process rather than content, learning is different; it has a broader focus, where understanding concepts is more important than the internalization of masses of detail. The PBL approach is based on active learning as opposed to passive learning, as shown in Figure 2.

Figure 2: Problem-Based Learning Cycle

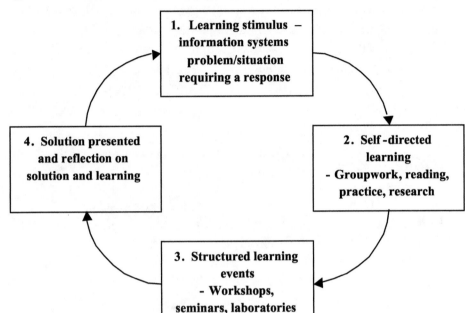

USING PROBLEM-BASED LEARNING

This section discusses course structure and organization of a PBL approach adopted by the authors in an introductory systems analysis and design course. The course has been offered using PBL on one of the smaller campuses of Victoria University in Melbourne, Australia. Numbers in the course are likely to be less than 60 students most semesters. The course consists of a scheduled two-hour tutorial/workshop, one-hour lecture and one-hour group meeting time. The schedule of lectures, tutorials and group meetings is spread across the week where possible, to promote opportunities for students to meet regularly and spread their learning interactions across the week. A small campus timetable permits this, though it may require additional effort to achieve on larger campuses and with large student numbers.

Learning Group Formation

A number of approaches to group formation have been tried. The groups in the past were rearranged for each problem and were self-selecting. This proved a difficulty as academically stronger students or friends tend to form groups. We have found it best if tutors assign students into groups based on their academic performance in the previous semester. The groups consist of five members, though larger groups can be formed. This number allows for a balance of abilities and breadth of learning across groups. Instructing groups that they are a learning group and will be together for the whole semester engenders a greater bonding relationship among students within a group. This also allows a range of group processes to arise and to mature. There is an attempt made to ensure gender balance within the groups.

Learning Resources

Students are provided with access to computer learning resources such as the Internet, intranet, self-assessment tests, discussion server and e-mail. Students complete weekly planning sheets to guide their learning for the coming week. The students submit weekly reflective learning diaries, where they record their reflections on their learning and activities. The keeping of meeting agendas and minutes is also encouraged.

Table 3: Learning Resources

Learning Resources	
On-line resources	Course Web site
	Internet access
	Intranet access
	Self-assessment tests
	Discussion server
	E-mail
	Planning sheets
	Reflective learning diaries
	Group meeting and agenda templates
Traditional resources	Textbooks
	Libraries
	Notes
	Workshops
	Laboratories
	Seminars
	Lecturers and tutors

Responsibilities

Students' expectations of the course, their tutor and their own responsibilities are discussed and agreed to between the students and the tutor in the first tutorial. These are placed on the course Web site. When issues arise, such as lateness or attendance, the expectations are used to remind individuals of the responsibilities they undertook. Through discussion of expectations, it is felt there is "ownership" by students of these expectations, rather than being tutor imposed.

Problem Solving

Students often need assistance in improving their problem-solving skills. To assist in their understanding and tackling of a problem, we use a "Know Need Do Assign (KNDA)" planning sheet as shown in Figure 3, with columns headed: "What we already Know," "What we Need to know," "What tasks we need to Do," and "Who the task is Assigned to." This sheet was refined from a KWL (Know, Want, Learned) Chart in Barell (1998, p.35) and a KND (Know, Need, Do) Chart in Fogarty (1997, p.6). Using the sheet, students define the problem statement, goals and constraints. This also helps students in conceptualizing the scope (scoping) of the problem. Each group uses this sheet to assist their initial understanding and

Figure 3: Know Need Do Assign Planner

KNDA Planner Name(s): _____ Problem statement: Situation: Goal: Constraints:			
What we already Know	**What we Need to know**	**What tasks we need to Do**	**Task Assigned to**

determination of the problem. They translate their assigned learning tasks onto their own individual learning plan for the week(s).

Reflection and Evaluation

As stated in the ITiCSE Working Papers (Ellis et al., 1998, p.52b): "A key issue in professional development is to have the skills of self-evaluation and the ability to steer one's activities." Self-evaluation and reflection on the problem are achieved through the use of diary entries, presentation and discussion of the solution, discussion in the group or together as a class, and sometimes setting a written question for another group to answer. This question allows groups to demonstrate their learning. For example, to test the understanding of a transaction processing system, a group once asked another group, "What are three transactions that take place in a university?" Self-evaluation tests, consisting mainly of 20-30 short answer and multiple-choice content-based questions, are given to students weekly. Students report that these tests help to confirm their learning and provide pointers for follow-up learning if there is an identified knowledge gap.

Course Structure

Table 4 shows the schedule of problems over the typical 13 weeks of a semester. Large problems, which we termed "Major Problems," involve a number of weeks. Smaller problems are completed in one week. The first small problem is used to introduce the PBL approach to students. The smaller problems are interspersed between the large problems to reinforce the learning approach and overcome the mid-semester motivational lapse that students often seem to experience.

Table 4 : Schedule of Problems

Week	Problem	Nature of the problem
1	Problem 1	Understanding systems and roles
2 - 5*	Major Problem 1	SDLC, analysis modeling techniques
6	Problem 2	Feasibility and project selection
7	Problem 3	Data gathering
8	Problem 4	Joint application design
8-13**	Major Problem 2	Methodology, requirements specification and prototyping
*Week 5 includes a presentation to the class and reflection on the problem. **Week 13 involves a presentation to the class and the client, followed by reflection on the problem		

Role of the Tutor

After reflection and discussion of the solution to a problem, a "mini" lecture is often given, using the one-hour lecture time when it is felt another perspective or insight could add to the students' new knowledge. The "mini" lectures last for 10 to 25 minutes. By attending mini lectures after reflection, the students seem to be more attentive than in normal lectures and appear to be able to relate the short lecture content to the problem and learning that has taken place. There appears to be more relevance for the students in the "mini-lecture" approach than in "lecture first followed by practice." The lecture time is also used to present material regarding topics that relate to the student questions or comments that were made in the previous week's submitted planning sheets and diaries. There is a responsibility on the part of lecturers and tutors to understand the learning needs of the students and to be prepared to give impromptu "mini" lectures. The students also ask for skills direction and explanations/discussion of concepts they have difficulty understanding. On occasion, short extra skill sessions are provided to accommodate specific skills acquisition that students request. In PBL there is a new role for tutors as facilitators or guides to empower students. Russell et al. (1994, p.59) suggest "educators are therefore required to implement strategies which promote self-directed learning skills, are conducive to students' construction of knowledge and promote reasoning skills."

Trial and Introduction of PBL

The PBL approach in this introductory systems analysis and design course at Victoria University was initially trialed with a small group of students and compared against a group undertaking the normal approach. This trial allowed the authors to become familiar with PBL and learn how it could be used within an information systems subject. This trial laid the foundation for changeover to a wholly PBL approach in the course.

Information on students' perspectives of the PBL trial and the subsequent change to the subject using PBL were gathered through observation, question-naires, interviews, focus group and diaries. Bentley et al. (1999a, 1999b, 2000) have reported the benefits, issues and student perceptions of the use of PBL. The benefits and issues are discussed in the following sections.

Benefits

Benefits perceived by both staff and students (Bentley et al., 1999a, 2000) include:
- increased motivation;
- improved problem solving;

- improved time management;
- improved self directed learning skills;
- improved research skills;
- improved group work skills.

Students have suggested that the problems we use are realistic and relevant to the work of an information systems professional. We have not found any appreciable differences between students' assignment and exam performances between teaching the course in the traditional mode than through the use of PBL. PBL can provide other benefits as well. For example, a tutor who worked with the same group of students in a programming subject and later in a PBL setting noted the increased enthusiasm and motivation shown by students in their group work and approach to learning.

In improving self-evaluation skills, students may for the first time in a course have to write down their reflections on their learning in diaries. Diary entries written earlier in the semester suggest a need to spend greater time on reading and practice and to consider the number of hours spent on the subject. As reflected in their diaries, many students seem to realize the necessary commitment required for effective tertiary study.

Issues

A number of issues arise in the use of the PBL approach as a teaching strategy. Major issues include:
- PBL is a new approach to learning for students and faculty;
- teaching space suitability;
- group work skills;
- assessment of process skills;
- attendance;
- facilitation and small-group teaching;
- time;
- developing problems for students.

Other issues arising from the student perspective (Bentley et al., 1999b, 2000) include:
- a focus on factual knowledge;
- problems with group work participation;
- weaker students requiring more direction;
- preparation and motivation for PBL.

New Approach to Learning

PBL is a new approach to learning for our students. The first four weeks emphasise teaching *about* the PBL approach. We have observed that students have to *unlearn* their prior conceptions of learning. Some students, especially the weaker ones, find responsibility for self-learning difficult and require extra tutor time for directing what they need to do and constant prodding to achieve a satisfactory solution to the problem. These students require reassurance that they are learning content as well as learning process skills as a valuable component of the course. PBL is being introduced into the second semester of the first year which may mean that students have already framed their perceptions of university education by undertaking subjects in their first semester that consist solely of traditional lectures and tutorials. It might be better to introduce PBL, or elements of PBL, in semester one for commencing students so that they can perceive that university education is different from secondary school and not necessarily focused on content learning. Hence, students will be able to develop their self-directed learning and collaborative skills gradually throughout their first year.

Physical Environment

Consideration has to be given to the design of teaching space to support PBL. Group work in computer laboratories caused difficulty, as the room arrangement often does not allow for students to interact well as a group. There are few tables where students can sit around for face-to-face discussions. Some groups initially sit in a line facing computers. When this is observed, the tutor and students pause to find a means of creating an effective meeting space that encourages face-to-face interaction. This situation provides an opportunity to discuss the dynamics of meeting as a group. Subsequently groups have been observed to employ better meeting tactics. Fortunately there are often free classrooms available opposite the computer laboratory that the students can utilize. This extra classroom space is not a resource that many universities can usually sustain. When adopting PBL as a major teaching approach, then, consideration must be given to investing in teaching space and arrangements to support PBL group work.

Learning Groups

A number of important issues regarding group work arise, such as leadership, group composition and meeting times. Group leadership is addressed through the use of a rotating chair or team leader on a weekly basis, which ensured that everyone in the group had at least two weeks as leader. The leader chairs meetings, acts as a communication facilitator and checks other team members' progress during the week.

Formation and composition of groups can have an impact on a student's learning. Teacher selection of semester-long learning groups, based on their demonstrated academic strength in the course from first semester, gives better balance to the groups than a self-selection approach by the students. We have seen that teacher-selected groups allow peer pressure from motivated, higher-achieving students to encourage other students to achieve the tasks required. However, at times this results in students having to resolve conflict, decide on the level of the task suitable for each person and face real team issues about contribution similar to those encountered in the professional work-place. It is felt that putting together a longer-term group fosters a team environment, though the problem of students moving between courses in the first four weeks of semester has to be addressed if groups are likely to shrink or be augmented by new members. This is not usually a problem on a smaller campus, but would need to be addressed in subjects with large enrollments.

Participation

Attendance, both at class contact time and at group meetings, seems to be crucial in our format of PBL. Marks are awarded to provide an incentive for students to maintain attendance and participation. Poor attendance by students at meetings outside of regular class times is often a major group work problem reported by students. The timetabling of an unsupervised laboratory hour for students to meet provides students with a common time and little excuse not to attend.

Time

The issue of time and group work arises for students. They are concerned about the time they spend on preparation, meeting time, reading and completing plans and diaries. This has an impact on students, as many are employed for more than 15 hours per week in order to provide for their university education. The completion of what students see as administrative tasks appears to encroach on their time. One student commented, "Administration takes more time than the learning." The tools to aid and reflect on learning are seen by some students to be a burden that is not encountered in other subjects. These comments are addressed in lecture time to the whole class and help to reinforce the change in learning they are encountering and the need for them to develop new skills in "learning how to learn."

Developing PBL Problems

Setting and developing the PBL problem is an issue for faculty. There has been uncertainty about the nature of the problems set as to whether the problems were PBL or represented situation-based learning (Russell et al., 1994, p.61). It is felt that many of the problems perhaps lay somewhere between the two. McCracken and Waters (1999) compare PBL problems in software engineering to medical school problems, suggesting there is a potentially significant difference between the problems in the two fields. Medical problems tend to be shorter, with solutions consisting of a diagnosis and proposed treatment, so the students can then move onto the next problem. In software engineering, however, there are deliverables developed over a longer period of time and the assessment of these is a substantial part of a student's grade. Explicitly setting the learning objectives for each problem assists in scoping the problem, so that solutions and learning are achieved within the time limitations imposed by the course. Curriculum development time needs to be given to source problems and write them up as PBL cases ready for tutors. Fully developed cases include the learning objectives and suggested strategies for tutors to guide and assist students. The students themselves only see the problems. The problems need to be authentic and presented as such, rather than being task-based or project-based. With wider discussion on the development of problems in the information systems education community, examples of appropriate problems may help to develop a norm for problems that lend themselves to PBL.

Weak Students

Weak students, and in some cases weak groups of students, can be left behind and are likely to exhibit only surface learning. These groups require extra encouragement and guidance, perhaps by external rewards, whereas deep learners may be intrinsically motivated (Conrick et al., 1994, p.250). In PBL it is important to identify weaker students and to provide them with encouragement and feedback. Some students in our course clearly indicate that PBL is not an approach that suits them. They want more direction and structure imposed on their learning. Often these students decide to undertake Technical and Further Education courses rather than higher education studies. The weakest students strongly suggest that they preferred directed learning. An unintended consequence of PBL may be a self-realization that assists students in identifying the approach to learning that suits them and whether they wish to be professionals or technicians in their computing career path.

INTRODUCTION OF PROBLEM-BASED LEARNING INTO THE CURRICULUM

There is no single model of PBL. A spectrum of approaches exists. All PBL approaches share the common principle that learning derives from a real-world problem. A fully integrated curriculum based on PBL would be inter-disciplinary and level, rather than course based. Most attempts at PBL commence with its use in individual courses, often the only such course in a given curriculum.

Advice is offered to those contemplating adoption of PBL on three major aspects. First, support and resources required for introducing PBL:

- Attend workshops on PBL (see Appendix A).
- Use professional contacts to assist in developing real-world problems for students.
- Gain support from management and others as you will need to negotiate and discuss timetabling changes, funding and staffing.
- Provide a leadership and role model for other staff to follow.
- Be flexible. You need to allow some space in the course in order to experiment with PBL.
- Be prepared to commit time to introducing PBL. There is little in the way of materials for PBL in information systems, so developmental time and resources will be needed.
- Seek support to appoint a PBL Coordinator. This person can assist others in adopting PBL and provide training for tutors.
- Visit the PBL links in Appendix A.
- Join a PBL discussion list (see Appendix A).

Second, be aware of possible PBL implementation problems. These include:

- Reluctance to change and experiment in the faculty.
- Faculty management is primarily interested in efficiency of teaching.
- Current culture in many faculties seems to discourage student-centered learning.
- Costs in setting up, especially in extra curriculum time and effort to develop problems. A large amount of staff time is needed to initially implement PBL ("time consuming but in a more creative way").
- PBL may need a department/school to have a lower tutor-to-student ratio.
- Lecturers need to change their role in PBL from expert to coach. This may be difficult for some people.
- PBL can be resource and time intensive, both for students and staff.
- Staff resistance: This is a mind change required for some faculty. Staff may have to rethink their teaching and learning approaches.

Third, to be an effective PBL tutor, information systems faculty require the following skills:

- High level of interpersonal skill.
- Ability to a manage a group.
- Good facilitation, able to facilitate learning in small groups.
- Recognition that they are no longer a performer.
- Ability to assess learning, not just retention of knowledge.
- Ability to monitor progress of both individual students and groups.
- Ability to help students reflect.
- Ability to create a positive learning atmosphere.
- Being a role model.
- Ability to regard students more as junior colleagues than as dependents.

CONCLUSION

Many, if not most, information systems graduates will be employed in a project environment. Information systems curriculum is often delivered in small, byte-sized, semester-long "chunks" that resemble buckets containing defined areas of knowledge suitable for related study. In each course, the academic is a *hurdle* to be gotten over by the students as they move "forward" toward a passing result. If they repeat this exercise successfully enough times, once per required course, we say to them "Now, you can go and integrate all of this," graduate them with their degrees and send them out to meet their fate. How well, though, do skill at study, passing examinations and producing student projects prepare information systems students for work in the professional project environment? Would they not be better served by working in a project environment, learning and practicing how to identify and structure problems, frame questions, locate information and resources, work with and serve clients and colleagues, to ascertain and locate the knowledge and resources needed, and to successfully plan and complete projects, from the beginning and throughout their period of study?

This chapter has presented a simplified model for development of a "compleat information systems graduate" to assist educators in developing curriculum content and choosing a learning approach. It is suggested this may be achieved through changing from a teacher-enabled to a student-enabled learning approach using PBL. This approach is in its infancy in the information systems discipline and is far from the fully integrated PBL curriculum used in many medical and law schools. The PBL approach is a good match with the realities of professional information systems practice by providing a closer link between the learning and work environment.

APPENDIX A: PROBLEM-BASED LEARNING RESOURCES AND LINKS

General PBL resources and information on the Web.

Adelaide University, Australia, PBL experiences -
  http://www.acue.adelaide.edu.au/leap/focus/pbl/
  http://www.acue.adelaide.edu.au/leap/leapinto/pbl/index.html
Illinois Maths and Science Academy, Centre for Problem Based Learning -
  http://www.imsa.edu/team/cpbl/problem.html
Maastricht University PBL - *http://www.unimaas.nl/~PBL/*
McMaster University, Donald Woods PBL Web pages -
  http://chemeng.mcmaster.ca/pbl/pbl.htm
Monash University, Australia, PBL in Civil Engineering. PBL information, resources, links to PBL conferences, institutions and discussion
  lists - *http://cleo.eng.monash.edu.au/teaching/pbl-list/*
Southern Illinois University School of Medicine PBL Initiative -
  http://www.pbli.org/
University of Delaware, PBL syllabi, sample problems and exams -
  http://www.udel.edu/pbl/
University of Newcastle, Australia, Problem-Based Learning Assessment
  and Research Centre (PROBLARC) at -
  http://www.newcastle.edu.au/services/iesd/learndevelop/problarc/

PBL Discussion Lists

UD-PBL-Undergrad list at the University of Delaware -
  http://www.udel.edu/pbl/ud-pbl-undergrad.html
PBL-LIST at Monash University -
  http://cleo.eng.monash.edu.au/teaching/pbl-list/pbl-list.htm
IMSACPBL-L (for grades K-16) -
  http://www.imsa.edu/team/cpbl/contact/listserv.html

PBL Workshops

Workshops are conducted by a number of universities with PBL centers, for
  example PROBLARC in Australia and Southern Illinois University in
  the USA.
Southern Illinois University School of Medicine workshops -
  http://www.pbli.org/workshops/index.htm
PROBLARC workshops
  http://www.newcastle.edu.au/services/iesd/learndevelop/problarc/
  wshop.htm

REFERENCES

Aldred, S.E., Aldred, M.J., Dick, B. and Walsh, L.J. (1997). *The Direct and Indirect Costs of Implementing Problem-based Learning into Traditional Professional Courses within Universities*. Canberra: Australian Government Publishing Service.

Barell, J. (1998). *PBL: An inquiry approach*. Australia: Hawker Brownlow Education,.

Bentley, J. F., Lowry, G. R., and Sandy, G. (1999a). Towards the compleat information systems graduate: A problem-based learning approach. In Hope, B., and Yoong, P., (Eds.), *Proceedings of the 10th Australasian Conference on Information Systems (ACIS)*, 1. Victoria University, Wellington, New Zealand: School of Communications and Information Management, December, 65-75.

Bentley, J. F., Lowry, G. R., and Sandy, G. (1999b). Initiatives in information systems: Matching learning to professional practice. In *Implementing Problem-Based Learning: Proceedings of the 1st Asia Pacific Conference on Problem-based Learning*, Hong Kong, December 9-11, 111-117.

Bentley, J., Sandy, G., and Lowry, G. R. (2000). Introduction to business systems development: Students' perspective of a problem-based learning approach. In *Proceedings of Information Systems Education Conference (ISECON)*. Philidelphia, USA.

Bloom, B. (Ed.) (1956). *Taxonomy of Educational Objectives: The Classification of Educational Goals, Handbook 1: Cognitive Domain*. London, UK: Longmans.

Boud, D. and Feletti. G. (Eds.) (1991). *The Challenge of Problem-Based Learning*. New York: St. Martin's Press.

Conrick, M. (1994). Problem-based learning: Managing students transitions. In Chen, S. E., Cowdroy, R. M., Kingsland, A. J. and Ostwald, M. J. (Eds.), *Reflections on Problem-Bases Learning*. Sydney, Australia: Australian Problem-Based Learning Network, 237-255.

Ellis, A., Carswell, L., Bernat, A., Deveaux, D., Frison, P., Meisalo, V., Meyer, J., Nulden, U., Rigelj, J. and Tarhio, J. (1998). Resources, tools, and techniques for problem-based learning in computing: Report of the ITiCSE'98 working group on problem-based learning. In *Working Group Reports of the 3rd Annual SIGCSE/SIGCUE ITiCSE Conference on Integrating Technology into Computer Science Education*, 30, Dublin, Ireland, 45b-60b.

Fogarty, R. (1997). *Problem-based learning and other curriculum models for the multiple intelligences classroom*. Australia: Hawker Brownlow Education.

Greening, T. (1998). Scaffolding for success in PBL. *Med Educ Online [serial online]*. http://www.utmb.edu/meo/

Little, P. (1999). *Introduction to Problem-Based Learning*. Workshop conducted at Victoria University of Technology, by PROBLARC, The University of Newcastle, Australia, February 26.

Lowry, G. R. and Doroshenko, E. E. (1996). Object-orientation in software engineering education: Integrating concepts, methodology development models, and CASE in the curriculum. In *Proceedings of the International Conference on Software Engineering: Education and Practice*. Dunedin, New Zealand: IEEE Computer Society Press, 336-343.

Lowry, G. R. and Morgan, G. W. (1996). Transition to object orientation in software engineering education. In *Proceedings of the International Conference on Software Engineering: Education and Practice*. Dunedin, New Zealand: IEEE Computer Society Press, 506-510.

McCracken, M. and Waters, R. (1999). Why? When an otherwise successful intervention fails. In *Proceedings of the 4th Annual SIGCSE/SIGCUE Conference on Innovation and Technology in Computer Science Education*, Krakow, Poland, 9-12.

Mulder, M. (1998). The NSF Project (1995-1998). *Presentation to Association for Information Systems (AIS)*, Baltimore, Maryland, August 16, unpublished.

National Office for the Information Economy (NOIE). (1998). *Skill Shortages in Australia's IT&T Industries*. http://www.noie.gov.au/projects/ecommerce/skills/discpaper_dec98/skills.html.

National Office for the Information Economy (NOIE). (1999). *International IT&T Skills Situation and Government Responses - Background Paper*. http://www.noie.gov.au/projects/ecommerce/skills/paper_internat_backgrnd_99.htm

Orr, J. and von Hellens, L. (2000). *Skill Requirements of IT&T Professionals and Graduates*. ACM SIGCPR, Chicago, 167-170.

Ralston, P. (1999). Uni bashing is cause for alarm. *The Australian*, April 20, 1999, 58.

Russell, A. L., Creedy, D. and Davis, J. (1994). The use of contract learning in PBL. In Chen, S. E., Cowdroy, R. M., Kingsland, A. J. and Ostwald, M. J. (Eds.), *Reflections on Problem-Based Learning*. Sydney, Australia: Australian Problem-Based Learning Network, 57-72.

Ryan, G., Toohey, S., and Hughes, C. (1996). The purpose, value and structure of the practicum in higher education: A literature review. *Higher Education*, 31, 355-377.

Saarinen-Rahiika, H. and Binkley, J.M. (1998). Problem-based learning in physical therapy: A review of the literature and overview of the McMaster University experience. *Physical Therapy*, 78, 195-207.

Savery, J. R., and Duffy, T. M. (1995). Problem-based learning: An instructional model and its constructivist framework. *Educational Technology*, 35(5), 31-38.

Turner, R. and Lowry, G. (1999). Reconciling the needs of new information systems graduates and their employers in small, developed countries. *South African Computer Journal,* 24 (November), 136-145.

Turner, R. and Lowry, G. (2000). Motivating and recruiting intending IS professionals: A study of what attracts IS students to prospective employment. *South African Computer Journal,* 26 (November), 132-137.

Turner, R. and Lowry, G. (2001). What attracts IS students to prospective employment: A study of students from three universities managing information technology in a global economy. In Khosrowpour, M., (Ed.), *Managing Information Technology in a Global Economy.* Hershey, USA. IRMA, pp. 448-452. ISBN: 1-930708-07-6.

Underwood, A. (1997). *The ACS Core Body of Knowledge for Information Technology Professionals.* http://www.acs.org.au/national/pospaper/bokpt1.htm.

Woods, D. R. (1996). *Problem-Based Learning: Helping Your Students Gain the Most from PBL.* (3rd ed). http://chemeng.mcmaster.ca/pbl/pbl.htm.

Woods, D. R. (1994). *Problem-Based Learning: How to Gain the Most from PBL.* (2nd ed). Waterdown, Canada: Donald R. Woods.

Woods, D. R. (1997). *Problem-Based Learning: Resources to Gain the Most from PBL.* (2nd ed). Waterdown, Canada: Donald R. Woods.

Section III

Impact of the Web on IT Technology

Chapter VII

Teaching or Technology: Who's Driving the Bandwagon?

Geoffrey C. Mitchell
Victoria University of Wellington, New Zealand

Beverley G. Hope
Victoria University of Wellington, New Zealand
and City University of Hong Kong, China

Fuelled by the increasing connectivity afforded by the Internet and the flexibility offered by Web technologies, the use of technology in education has become increasingly common. However, despite claims that the Web will revolutionise education, many attempts at Web-based education simply reinforce current 'poor' teaching practices or present more of the same disguised in updated packaging. We argue that this occurs because of differing pedagogical assumptions and a limited understanding of how flexible learning differs from traditional approaches. In particular, we argue that flexible learning demands an increased focus on constructivism and the sociological aspects of teaching and learning. This chapter presents two frameworks that situate our approach to flexible learning with respect to more traditional offerings and discusses the implications for educational technology design.

Copyright © 2002, Idea Group Publishing.

INTRODUCTION

Advances in information and communication technologies (ICT) in the last 20 years have had a significant social and economic impact (Adams & Warf, 1997). The diffusion of ICT includes not only the ubiquitousness of computers, but also increases in computing power, multi-media capability, and interconnectedness. The education sector has not been immune to the impact of these developments. While moves from purely synchronous to asynchronous delivery modes preceded the ascendance of ICT, ICT has been a major enabling factor in more recent shifts from broadcast (1-m) to interactive (m-m) interaction modes and from linear (textbook) to network (hypertext) information presentation.

The new technologies have both 'pull' and 'push' impacts on trends in flexible education. That is, they provide a solution to demands for more flexible forms of education and, because of their capabilities, also serve to increase demand. Informed, motivated students in the new competitive educational environment are demanding modes of learning that suit their individual needs.

Life-long education and global course offerings have led to increasingly diverse student populations at the higher education level. This diversity is apparent in demographic characteristics such as age, culture, prior education, work and life experience, and learning style. Meeting the needs of this diverse population requires greater flexibility in course delivery. This is difficult to achieve in a traditional, large-class, same-time/same-place teaching environment. Increased availability of computers and greater connectivity can overcome these difficulties. Regrettably, many Web-based educational implementations reinforce rather than replace inflexible teaching practices.

In this chapter, we argue that to maximise the benefits of educational technology, greater attention must given to the motives and planning behind the adoption of educational technology. We first discuss some of the underlying pedagogical assumptions of instructional choices from three perspectives: techno-logical determinism, psychological determinism, and sociological determinism. This forms the background to the development of a framework comparing 'emergent' teaching and learning practices with more traditional practices. Emergent, flexible forms of education are discussed in terms of their implications for educational technology design. We conclude by presenting some simple guidelines for the design of educational technology to support both teaching and learning practices.

BACKGROUND

Underlying the debate about the role of ICT in education are some basic pedagogical assumptions (Berman, 1992; Cowley, Scragg, & Baldwin, 1993; Harsim, 1990; Kozma & Johnston, 1991; Miller & Miller, 1999) and some theories about the relationship between technology and organizations (Eason 1988, 1993; Hirschheim, 1985; Keen, 1981; Leonard-Barton, 1988; Markus & Robey, 1988; Sproull & Goodman, 1990). Differing assumptions influence expectations and implementations. These assumptions can be grouped and summarised as three perspectives:

1.   technological determinism,
2.   psychological determinism, and
3.   social interactionism

Our review of these three perspectives is necessarily brief, and to that end simplified. However, an understanding of the underlying assumptions is essential for understanding approaches to and pitfalls of educational technology implementations.

Technological Determinism

Technological determinism assumes an intrinsic worth of technology and is optimistic about the effect ICT will have on education. Opportunities for adding value are considered to depend on the features and quality of the technology and on users' ability to recognize them (Campbell, 1996). In this respect, technological determinism takes a utopian view of technology.

The view discounts the impact of the context into which technology is introduced. Issues such as power and politics are ignored because teachers and students are considered rational beings, following logical decision-making processes toward shared goals (Markus & Robey, 1988). Technology is an "exogenous force which determines or strongly constrains the behaviour of individuals" (Markus & Robey, 1988, p.585). Consequently, the technology rather than the educational environment, the teacher, or the learner determines the success of an educational technology implementation.

Technological determinism in educational technology design results in an over-emphasis on hardware and technical capability and an under-emphasis on student needs and characteristics. The focus is on the technology in the confident belief that "If we build it, they will come."

Psychological Determinism

This pedagogical perspective is founded in cognitive processing theories of the 1970s and 80s. These theories focused attention on internal mental processes of learners and provided an alternative to the behaviourist approaches which had motivated earlier programmed instruction and computer-assisted instruction (Miller & Miller, 1999). However, like behaviourism, cognitive processing theories follow an objectivist paradigm in assuming that knowledge is independent of and external to the learner.

A belief in objective and external knowledge leads to teaching strategies in which the instructor (knowledge expert) structures and presents knowledge in such a way that the learner can accurately acquire it. The educational objective is to transmit knowledge. Successful learning is evidenced by the learner's ability to correctly reproduce this knowledge (Cronin, 1997; Jonassen, Davidson, Collins, Campbell, & Haag, 1995).

Educational technology design under this perspective focuses on the structure and presentation of knowledge. Information structure is likely to be strongly hierarchical, mirroring the reality of the external knowledge. Presentation considerations include giving attention to such issues as font size, use of colour, and use of images in the belief that these will aid attention and foster knowledge retention. Web-based instruction founded in this perspective can result in the placement of traditional course material onto static Web sites.

Social Interactionism

Social interactionism is founded in the constructivist paradigm. In contrast to psychological determinism, constructivism does not view knowledge as external to the learner. Rather, knowledge is internally constructed as the learner attributes meaning to experiences within their environment (Bereiter, 1990; Cronin, 1997; Jonassen et al., 1995). Purists under this view may hold that knowledge has no objective reality, while moderates may acknowledge an objective reality but one which can only be subjectively known (Miller & Miller, 1999).

Whatever the degree of a constructionist's belief, society cannot function on purely individualistic conceptualisations of knowledge. Some consensus of meaning must be obtained and this is achieved through social interaction with others (Heylighen, 1997). This social aspect of knowledge construction is often termed 'social interactionism,' because it necessitates interaction and collaboration.

The social interactionism perspective offers no prescriptions for educational success, since requirements depend on the particular skills, abilities, and needs of individuals. Nevertheless, several authors have produced lists of implications of constructivist theories for learning and teaching. Murphy (1997) reviews some of the major theorists and summarises the implications as:
1.   Multiple perspectives and representations of concepts and content are presented.
2.   Goals and objectives are derived by the learner or in negotiation with the teacher.
3.   Teachers serve as guides, monitors, coaches, tutors, and facilitators.
4.   Activities, tools, and environments encourage metacognition, self-analysis, self-regulation, self-reflection, and self-awareness.
5.   The learner plays a central role in mediating and controlling learning.

6. Learning environments, content, and tasks are relevant, realistic, and represent the natural complexities of the 'real world.'
7. Primary sources of data are used to provide authenticity and real-world complexity.
8. Knowledge construction and not reproduction is emphasized.
9. Knowledge construction takes place in individual contexts and through social negotiation, collaboration, and experience.
10. The learner's previous knowledge constructions, beliefs, and attitudes are considered in the knowledge construction process.
11. Problem-solving, higher-order thinking skills, and deep understanding are emphasized.
12. Errors provide opportunities for insight into learners' previous knowledge constructions.
13. Exploration is a favoured approach to encourage learners to seek knowledge independently and to manage the pursuit of their goals.
14. Learners are provided with the opportunity for apprenticeship learning in which there is an increasing complexity of tasks, skills, and knowledge acquisition.
15. Knowledge complexity is reflected in an emphasis on conceptual interrelatedness and interdisciplinary learning.
16. Collaborative and cooperative learning are favoured in order to expose the learner to alternative viewpoints.
17. Scaffolding is facilitated to help students perform just beyond the limits of their ability.
18. Assessment is authentic and interwoven with teaching.

These tactics can be difficult to achieve with large undergraduate classes. However the use of ICTs, and in particular Web technologies, can greatly facilitate their successful implementation.

The three perspectives outlined – technological determinism, psychological determinism, and social interactionism – show how differing assumptions affect education and educational technology design and implementation. Each perspective contributes something to our understanding, but given the prevalence of technological determinism and the entrenched practices derived from psychological determinism, it is important in our efforts to develop flexible education that we give consideration to the tactics arising from a social interactionism perspective.

TRADITIONAL AND EMERGENT
FORMS OF EDUCATION

According to Willis (1995), traditional education is characterised by the following:

- learning process is sequential and linear;
- planning is top down and systematic;
- course objectives guide all learning activities;
- teachers are perceived as experts with special knowledge;
- careful sequencing of teaching activities is important to learning predefined skill sets;
- the main goal is delivery of pre-selected knowledge;
- summative evaluation forms the basis of assessment; and
- objective results are critical.

Traditional education revolves around the group lecture, supplemented by tutorials or workshops designed to reinforce the material 'delivered' by the lecturer. Courses are often modularised with each week covering a new area of knowledge or skill set. Learning objectives are carefully spelled out in course outlines and clearly linked to the various pieces of assessment. Courses emphasize parity and are readily accepted by faculty steering committees and external moderators.

While well accepted and often successfully executed, this form of learning generally fails to engage students in deeper learning (Mills-Jones, 1999) or to deliver the problem-solving skills favored by employers (Bentley, Lowry, & Sandy, 1999). Some instructors, departments, or disciplines have adopted problem-based learning and active learning strategies to address these concerns. However, until recently the logistics of dealing with large student numbers have limited the opportunities in these areas.

To date, economic issues have commonly been the driving force behind flexible learning and educational technology initiatives. The drive to attract more students and the associated funding at least cost has seen schools and universities adopt Web-based systems which merely deliver traditional mail-based distance courses over the Internet. Under the banner of 'flexible education,' instructors have provided online access to lecture notes and in some cases streamed lectures using Web-based audio and video technology. Unfortunately, there has been limited evidence that this newer approach to material delivery has added any real value to students' learning experiences (Riddle, Nott, & Pearce, 1995; Vargo & Cragg, 1999). The nature of the course and the assumptions of student learning are fundamentally the same as those of traditional education. It has been argued that to achieve real added value, the use of educational technology must be adopted in conjunction with a more fundamental change in the nature of student learning activities (Lamp & Goodwin, 1999; Mak, 1995; Ramsden, 1992; Vargo & Cragg, 1999). This is true flexible learning.

Situating Flexible Learning

Sociological theory has for many years recognised the essentially dualistic nature of human activity systems. Parsons and Bales (1955) labelled these dimensions task and socio-emotional, reflecting the difference between the technical aspects of work-related structures and the social concerns of the human actors operating within those structures. The theory suggests that these activity dimensions are central to individual and group development of values and beliefs (Pfeffer, 1981), a perspective in keeping with the notion of education as developing new value and belief structures or schemata. Consequently, an argument can be made for the view that because education is an important context for personal interaction – second only to 'work' – it is central to the social creation of meaning.

Student development of shared knowledge and meaning can be described within the context of the two dimensions. Students interact at many levels to create, confirm, and recreate shared meanings. This shared meaning construction covers a much larger knowledge domain than has traditionally been considered in the context of academic learning. The broader definition is consistent with Brown's (1994) examination of outside school learning, Fensham's (1992) work on commonsense knowledge, and Resnick's (1987) exploration of learning and reasoning outside school. However,

Figure 1: Socio-Technical Domains of the Educational Context

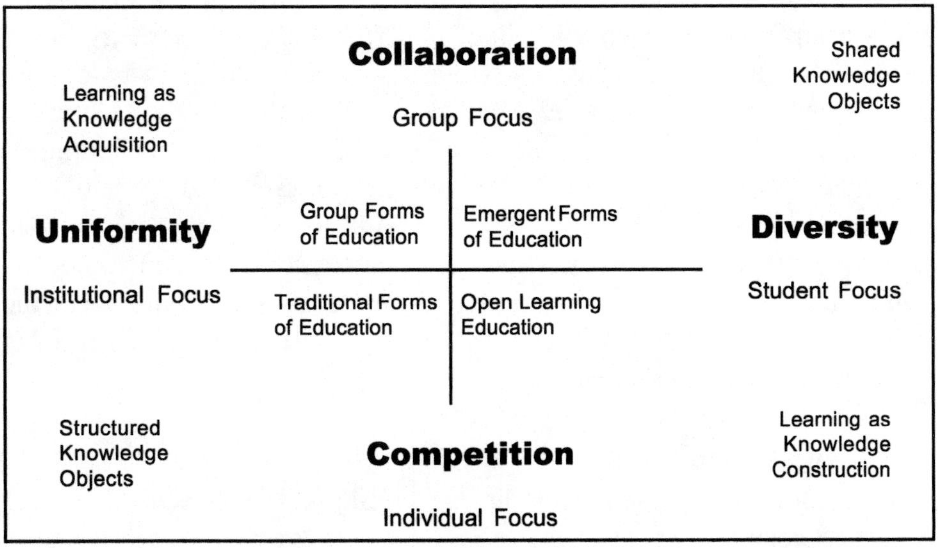

educational research has often identified these knowledge construction activities as separate functions taking place (1) inside and (2) outside of school. This view denies the role of social interaction within school in constructing shared meaning.

We situate approaches to education within this two-dimensional framework to highlight the differing perspectives and assumptions of teachers. Assumptions regarding the appropriate structure for education (task) can be thought of in terms of a uniformity-diversity dimension, while assumptions regarding appropriate student relations (socio-emotional) may be represented in a competition-collaboration dimension (Figure 1).

Uniformity-Diversity Dimension

The uniformity-diversity dimension distinguishes between those approaches to curriculum design that concentrate on singular notions of subject design, course delivery, and assessment, and those which emphasize flexible approaches. Uniformity of task is institutionally focused, that is, it concentrates on meeting university-wide standards or norms by offering the 'same' experience and assessment to all. Such programmes are readily accepted by boards of studies and external moderators. By contrast, diversity of task is student focused, offering different learning experiences for individual students. Diversity is often accepted in principle but less frequently practiced and less readily accepted. The uniformity-diversity distinction has received a great deal of attention in recent research into effective teaching and learning practices (Ramsden, 1992; Ramsden, Margetson, Martin, & Clarke, 1995).

Behind the current calls for more flexible teaching and learning practices lies the belief that teaching, particularly in higher education, needs to adopt a stronger student focus (Boyer, 1990; Brown, 1994). While academics who adopt a uniform approach talk about the 'right' teaching strategy, those interested in diversity deny the existence of any single 'right' way to teach (Bentley, Lowry, & Sandy, 1999; Mak, 1995; Mills-Jones, 1999). They argue instead that 'best' practices can only be based on an understanding of particular student learning needs. Thus uniformity is grounded in psychological determinism, while diversity is grounded in social interactionism. The growing pressure for diversity stems from a number of factors, including the diverse nature of the student population, the diverse needs of employers, the diversity of educational intentions, and the need to provide greater levels of teacher satisfaction with the educational process.

Competition-Collaboration Dimension

The competition-collaboration dimension reflects differing perceptions concerning the complex social interactions between students which are so important in developing shared knowledge objects. The extremes of this dimension are some-what multi-faceted in that they deal with the distinction between an individual focus (competition) and a group focus (collaboration), as well as dealing with the underlying assumptions regarding student attitudes to both their peers and teaching staff (Johnson & Johnson, 1989).

At the simplest level, the competitive aspect of the socio-economic dimension views students as a large collection of individual units (Bereiter, 1990), a perspective strongly evident in traditional educational activities. This view assumes students are motivated primarily by assessments, and that it is through assessments that students can be manipulated to receive planned educational objectives. By contrast, the collaborative end of the dimension views students as active constructors, rather than passive recipients of knowledge (Brown, 1994). The view assumes that learning is best facilitated through shared learning experiences among students as well as between students and teachers (Kushan, 1994). These views are often represented in notions like communities of learners (Brown, 1994), cooperative learning (Johnson & Johnson, 1989), and professional communities (Lieberman, 1992). Such views of the learning context and student interaction have strong implications for the appropriateness of current assessment tactics and the role students should play in them.

The intersection of these two dimensions results in four domains representing traditional education (uniformity/competition), group education (uniformity/collaboration), open learning (diversity/competition), and emergent forms of education (diversity/collaboration). The focus of this chapter is on the emergent forms of education resulting from an increased focus on diversity and collaboration.

Traditional Education

The uniformity-competition domain is strongly indicative of traditional forms of education. Within this domain, teaching is highly structured, singular in nature, and focused on rational teaching strategies in line with deterministic psychological theory. It is concerned with institutional procedures and practices and treats learning as a process of knowledge acquisition. Improvements in teaching strategies in this domain focus on finding better ways to deliver information or to motivate students to perform better in assessment exercises.

Group Education

The uniformity-collaboration domain is characterised by highly structured group forms of education. Learning is a process of knowledge acquisition, and the value of group interaction lies in the opportunity to reinforce the acquisition process. There is some conflict between the assumptions of the uniformity dimension and those of the collaborative dimension, evidenced by oft-repeated concerns regarding group assessments. These conflicts are responsible for a decrease in the popularity of group assessment activities in higher education.

Open Learning

The diversity-competition domain includes some current open learning activities. Students are still viewed as individual units, but more flexible teaching tactics are used. More recent open learning practices tend to treat learning as a process of knowledge construction rather than knowledge acquisition (Mayer, 1992), leading to emergent forms of education.

Emergent Forms of Education

We give the diversity-collaboration domain particular attention because we argue that this domain contains much of the current thinking regarding emergent forms of education. While the other three domains are largely representative of current teaching practice, this fourth domain is largely untested. We label this domain 'emergent forms of education.'

Our thinking behind labelling the diversity-collaboration domain 'emergent forms of education' is based on the definition of emergent as newly independent or becoming known as a result of inquiry (Sykes, 1982). Such a definition fits well with new conceptions of learning which focus on notions of inquiry and increased student responsibility for learning (Ramsden, 1992).

The domain views learning as shared knowledge construction, in keeping with notions of scholarship which emphasise a broad conception of knowledge gathering and dissemination as part of academic learning. The domain holds that the development of shared knowledge is best facilitated through collaborative student interaction (Staehr, 1993). There is also a strong focus on contextualising learning in the environment in which it finds its applicability (Friedman & Kahn, 1994). There is a focus on frequent interaction with students and an emphasis on students "organising the content and sequence of the teaching-learning process" (Riding & Cheema, 1991, p. 198). There is also a greater focus on encouraging divergent thinking and encouraging students to develop independence and responsibility for their learning engagement (Ramsden, 1992).

Overall, the domain is based on the assumption of education as a participative process in which there are very few absolutes and where both the content and the context of the learning process are open for negotiation between students and teachers. The underlying assumptions of this domain inform emergent educational practices and student expectations.

Characteristics of Emergent
Forms of Teaching and Learning

Emergent forms of education will expand educational technology beyond current conceptions of flexible delivery. We envisage a three-stage model that highlights the maturity of flexible learning initiatives, from flexible delivery through flexible interaction to flexible exploration (Figure 2). The majority of current educational technology implementations fit into the first stage and very few reach the third stage. Consequently, we represent the three levels as a pyramid, highlighting the present focus on lower level activities and the need to progress to higher level activities.

The *first stage*, flexible delivery, embraces the majority of current Web-based delivery efforts. The *second stage*, flexible interaction, supports a more cooperative form of flexible delivery by using Web and Internet technologies to enable students to interact outside traditional boundaries. Many course Web sites already utilise bulletin board and chat facilities. However, this is often an adjunct to the learning process and simply continues the information delivery notion. Movement from the first level to the second implies a more conscious utilization of synchronous and asynchronous communication devices to engage students in both lecturer-to-student and peer-to-peer investigations of problems being explored. The *third stage* is characterised by student exploration of concepts using Web and Internet technologies. At this stage, students can explore concepts in a non-linear, student-directed fashion. They can, in effect,

Figure 2: A Maturity Model of Flexible Learning Facilities

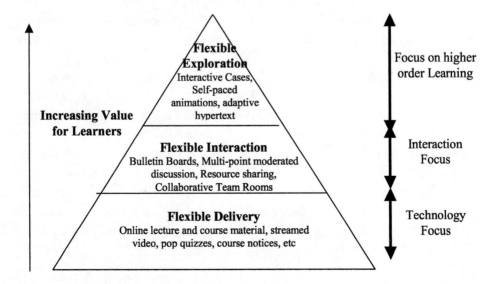

design their own learning experiences. For the most part, little effort has been expended at this level.

The emergent forms of education, situated in the diversity/collaboration quadrant, encompass the flexible interaction and flexible exploration stages. These are summarised next.

Flexible Interaction

If we accept that effective teaching and learning involve a process of developing and changing individuals' values and beliefs, enabling them to see the world from a different perspective, then it is appropriate to view the educational process as complicated layers of social interaction. But, no matter which definition of teaching and learning we subscribe to, a high level of student-student or student-teacher interaction is inevitable. The Web and the Internet offer a wide array of opportunities to support student interaction. The facilities range from the simple bulletin board, ICQ, IRC, and other generic chat facilities to the more sophisticated online tutorial facilities like Net Meeting, White Pine's Meeting Point and Class Point, and Lotus's Learning Space. Web-based courseware products like Web CT, TopClass, eCollege, and Blackboard's Course Site, which offer interactive facilities, are increasingly being used. However, recent surveys have shown that communication facilities alone will not deliver increased value to students or academics (Parry, Cockroft, Breton, Abernethy, & Gillies, 1999; Vargo & Cragg 1999). It is not the technology itself, but the way it is integrated into the learning process that creates value for students.

An example of an integrated learning process using Web and Internet technologies is the virtual collaborative team room. Traditional project groups often base themselves around a team room–a physical location where material related to the project can be grouped together in a work environment. As a collaboration tool, the physical team room is often highly effective. Group members are able to enter the room and begin working by themselves on a particular piece of the project, and then easily adapt themselves to working in groups should another member arrive. Further, due to its ability to develop project memory, the physical team room facilitates group members to pick up where another member left off on a particular task, thereby allowing for time-diverse collaboration. But the physical team room requires the co-location of all members of the project group. An alternative is to create a virtual team room utilising groupware applications. By locating the team room within an Internet-connected computer system, group members are able to participate in the project, and obtain many of the advantages of the physical team room even if they are in geographically diverse locations (Davis, Motteram, Crock & Mitchell, 1999).

Flexible Exploration

Flexible exploration focuses on students as active constructors of knowledge (Brown, 1994). Practically this is reflected in the widespread adoption of case-based and problem-based learning approaches to student learning (Bentley, Lowry & Sandy, 1999; Prawat, 1993). One of the perceived advantages of these methods is that they allow students to deal with complex but realistic situations, to gain experience in practical decision-making, and to develop their own models and approaches for dealing with unstructured problems (Bentley, Lowry, & Sandy, 1999). It is claimed that student recall and understanding of key concepts is improved when they are actively engaged in such realistic learning activities (Mills-Jones, 1999).

However, too often case and problem situations become overly sanitized and artificially abbreviated. The loss of richness in the material is commonly a product of tight teaching schedules or attempts to highlight the lecturer's perception of the important aspects of the case. As a result, students tend to approach the case as an exercise in re-organising the material to find the answer. This reduces both the experiential and realism aspects of the learning exercise (Bentley, Lowry & Sandy, 1999; Friedman & Kahn, 1994; Gallagher, 1994). The Internet and the Web offer a range of opportunities to address these concerns, particularly in terms of increased realism, improved student motivation, and an increased emphasis on self-directed student exploration of the material. At the simplest level multimedia facilities can enhance the visual aspect of case material. Audio and video tools can be used to upgrade case exhibits and break up the written components of the material. This is often enough to improve student motivation to explore the case material on their own or with peer groups.

In teaching cases specifically developed to take advantage of new technologies, communications tools and simple artificial intelligence agents can be used to encourage student interaction with the virtual actors in the case. Controlled interaction with artificial organisational units like steering committees or management groups can be supported allowing students to explore the material in ways not originally conceived by the lecturer. The student experience can also be enhanced by the ability to contextualise the case material within the wider context of resources available on the Web. These features all add to the complexity and realism of the problem being explored.

Another initiative that supports student-centered, self-paced learning is the use of adaptive hypertext. Technologies like XML and knowledge mapping concepts like Walden's Paths can allow students to direct their own exploration of the course material (Shipman, Furuta, Brenner, Chung, &

Hsieh, 1998). By mapping all the concepts covered in online course material using XML, it is possible to allow students to retrieve information based on search criteria they establish rather than the week-by-week delivery that currently defines information retrieval.

Use of online feedback can address an often-cited criticism of flexible exploration, namely, a lack of feedback and direction provided to students. For example, results of online quizzes can provide a nominal value that represents how well a student understands particular concepts. Based on this feedback, course material relevant to identified 'problem' areas can be collected by meta-crawler programs and delivered to the student. Given that many university undergraduate courses are reaching class sizes of over 300 students, and that tutorial classes often do not benefit shy or less self-assured students, adaptive hypertext as a supplement to traditional methods can potentially offer a more effective feedback mechanism than traditional learning environments can offer alone.

IMPLICATIONS FOR EDUCATIONAL TECHNOLOGY DESIGN

Ramsden (1992) suggests six key principles for effective teaching in higher education, many of which are in keeping with the emergent forms of education.

- raising student interest and providing clear explanation of concepts;
- having genuine concern and respect for students and student learning;
- providing appropriate assessment mechanisms and useful feedback;
- setting a clear goal and providing an intellectual challenge;
- encouraging student independence and control over learning engagement; and
- learning from students.

These principles all have a strong student focus and imply the need for a flexible approach to subject design and execution. In addition, the principles depend on a strong collaborative interaction among students and between students and teachers.

Brophy and Alleman (1991) pose a series of relevant comments for curriculum developers that are also important in the development of educational technology. They state that curriculum developers should pay attention to some very fundamental questions:

- What are the intended functions of activities within various types of curricula, and what is known about the mechanisms through which they perform these functions (if they do)?
- What is it about ideal activities that make them so good?
- What are some common faults that limit the value of less ideal activities?

- What principles should be followed by curriculum developers in designing activities and by teachers in implementing them with students?
  (Brophy & Alleman, 1991, p. 10)

By adapting similar principles, and by taking into consideration the issues outlined in the previous discussion on emergent teaching practices, it is possible to outline a number of educational technology design principles. The following may be considered good principles for the design and development of effective educational technology systems to support emergent teaching practices.

1. Support appropriate curriculum goals and objectives

Educational technology needs to be developed to support educationally derived objectives and goals. This necessitates detailed exploration of how each technological mechanism supports learning in specific higher education contexts. Well-designed educational technology should allow multiple teaching goals to be realised. For a system to be widely useful and broadly accepted, it needs to fulfill a variety of educational objectives and student needs. The most important design question for educational technology is probably whether the system should be developed at all.

2. Contextualise learning

The use of new technologies offers the opportunity to contextualise learning in ways not previously possible. The use of on-line systems and simulators allow for student exploration of knowledge domains within a framework resembling actual practice. It is particularly important that technology does not de-contextualise learning by conveying mixed messages.

3. Create communities of self-directed learners

Interactive communication facilities are of value only when they create communities of learners and not when they simply provide additional avenues of information delivery. Well-designed technology-based systems also ensure that students progress in their learning activities, while allowing them to broaden their understanding and explore tangents if they wish. One of the strengths of emergent educational technology is its ability to broaden students' conceptions of the world. Pop quizzes are useful tools, animations can engage, and discussion forums can support communication, but none of these by themselves will promote learning. Good educational technology should not promote rote learning but encourage exploration, thinking, and problem solving.

4. Encourage and facilitate open discourse

Many existing technology-based educational systems are designed simply to provide a more flexible method for the delivery of existing teaching material. But traditional face-to-face delivery allows students the chance for social learning not automatically present in technology-mediated delivery. In traditional delivery, students meet together in class and in small groups as they walk to or from class. In these groups students may talk about such things as their reactions to material offered, assignments, and workload coping strategies. New educational technology also needs to provide facilities which encourage and allow interaction between students and between students and teachers. This discourse should, where possible, be removed from the limitations of time and locational constraints.

5. Provide appropriate feedback channels

Much of the existing educational technology allows only for one-way communication of material. It is important, in conjunction with the previous principle, that emergent systems allow for teachers to support and monitor student learning through the provision of appropriate feedback. This feedback is an important aspect of any students' performance and perceived self-efficacy (which in turn affects future performance). The feedback should go beyond traditional communication of results and assessment feedback and incorporate more feedback that supports, challenges, and motivates students.

6. *Support teaching and learning needs*

When educational technology is developed from a purely pedagogical perspective, it is likely to support student needs. However, it is important that emergent systems also support teachers' needs. A system's 'success' is highly questionable if it creates extra burdens for teachers or removes those aspects of teaching that are considered enjoyable. A good system should enhance a teacher's participation with students and support other teaching-related issues like satisfaction, motivation, and career concerns.

7. **Engage Students**

Badly designed educational technology can create a gulf between students and teachers, increasing student alienation and reducing their learning capacity. In these circumstances, it is likely that there will be resistance to technology use. It is possible to overcome fear of technology by supporting a variety of learning experiences and by providing choice in learning approaches. Well-designed systems should enhance a student's perception of self-efficacy and not detract from it. Educational technology should highlight a teacher's interest in their students and not create a perception of distance.

8. **Be accurate, adaptable and flexible**

Good educational systems can be costly, complex, and difficult to create and maintain. It is important that anticipated benefits are not negated by poor instructions, simple errors, and technology failings. Systems must be adaptable so that teachers can ensure currency of the material provided. It is also important that they be flexible to allow teachers to adapt the teaching practices to suit different student groupings and learners to adapt them to their individual learning styles. The need for structure in a technology must not remove a student's ability to self-regulate their learning experience.

CONCLUSION

The Web can have a profound effect on course delivery, particularly in relation to developing delivery mechanisms capable of meeting the individual needs of diverse student cohorts. As students become more informed and ICT capabilities develop, there is a need for better-designed educational technology that supports teaching practices rather than acts as a billboard advertising technological wonder.

Our thinking on flexible learning is influenced by our belief in a multi-staged approach to flexible learning, built on the foundations of constructivism and social interactionism. The advantages of constructivism lie in the emphasis on learning as a process of sense-making through experience. Learning under these modes is active and reflective techniques which have been proven to increase understanding and retention. Good ICTs offer many opportunities for students to become active, reflective, self-directed learners. The goals of education technology must be to support truly flexible forms of education, improve opportunities for learning, and cater for different learning styles. Distance learning in disguise will not result in long-term benefits to students and teachers. The key lesson we have learned from our development and assessment of educational technology over the last few years is that educational technology produced without a clear pedagogical foundation rarely results in improved student learning. Finally, the point must be made that educational technology should not be considered a focal point for educational reform but rather a resource to be integrated into a wide repertoire of educational resources.

REFERENCES

Adams, P.C., & Warf. B. (1997). Cyberspace and geographical space. *Geographical Review*, 87(2), 139-145.

Bentley, J. F., Lowry, G.R., & Sandy, G.A. (1999). Towards the compleat information systems graduate: A problem-based learning approach. In Hope, B. G., and Yoong, P., (Eds.), *Proceedings of the 10th Australasian Conference on Information Systems*, 65-75. Wellington, New Zealand, 1-3 December.

Bereiter, C. (1990). Aspects of an educational learning theory. *Review of Educational Research,* 60(4), 603-624.

Berman, A. M. (1992). Class discussion by computer: A case study. *SIGCSE Bulletin,* 24(1), 97-101.

Boyer, E. L. (1990). *Scholarship Reconsidered: Priorities of the Professoriate*. Princeton, NJ: The Carnegie Foundation for the Advancement of Teaching.

Brophy, J., & Alleman, J. (1991). Activities as instructional tools: A framework for analysis and evaluation. *Educational Researcher,* 20(4), 9-23.

Brown, A. L. (1994). The advancement of learning. *Educational Researcher,* 23(8), 4-12.

Campbell, H. (1996). A social interactionist perspective on computer implementation. *Journal of the American Planning Association*, 62(1), 99-107.

Cowley, B., Scragg, G., & Baldwin, D. (1993). Gateway Laboratories: Integrated, interactive learning modules. *SIGCSE Bulletin,* 25(1), 180-183.

Cronin, P. (1997). *Learning and Assessment of Instruction*. Web document. http://ww.cogsci.edu.ac.uk/~paulus/Work/Vranded/litconsa.htm Last accessed on 8 March 2001.

Davis, T. J., Motteram, A., Crock, M., & Mitchell, G. (1999). Recreating a university: New approaches to learning, teaching and technology. *Tenth International Conference on College Teaching and Learning*, Jacksonville, Florida.

Eason, K. D. (1988). *Information Technology and Organisational Change*. London: Taylor and Francis.

Eason, K. D. (1993). Gaining user and organizational acceptance for advance information systems. In Massser, I., and Onsrud, H.J., (Eds.), *Diffusion and Use of Geographic Information Technologies* (pp. 27-44). Dordrecht: Kluwer.

Fensham, P. J. (1992). Commonsense knowledge: A challenge to research. *Australian Educational Researcher,* 20(1), 1-19.

Friedman, B., & Kahn, P.H. (1994). Educating computer scientists: Linking the social and the technical. *Communications of the ACM,* 37(1), 65-70.

Gallagher, J. J. (1994). Teaching and learning: New models. *Annual Review of Psychology,* 45, 171-195.

Harsim, L. M. (1990). Online education: An environment for collaboration and intellectual amplification. In Harrasim, L., and Turoff, M., (Eds.), *Online Education: Perspectives on a New Environment,* 39-64. New York: Praeger Publishers.

Heylighen, F. (1997). *Epistemological Constructivism.* Principia Cybernetica Web Document. http://pespmc1.vub.ac.be/construc.html (last accessed on March 8, 2001).

Hirschheim, R.A. (1985). *Office automation: A Social and Organisational Perspective.* Chichester: John Wiley.

Johnson, D., & Johnson, R. (1989). *Cooperation and Competition: Theory and Research.* Minnesota: Interaction Book Co.

Jonassen, D., Davidson, M., Collins, M., Campbell, J., & Haag, B. (1995). Constructivism and computer-mediated communication in distance education. *The American Journal of Distance Education,* 9(2), 7-26.

Keen, P. G. W. (1981). Information systems and organisational change. *Communications of the ACM,* 24(1), 24-33.

Kozma, R. B., & Johnston, J. (1991). The technological revolution comes to the classroom. *Changes,* 23(1), 10-23.

Kushan, B. (1994). Preparing programming teachers. *SIGCSE Bulletin,* 26(1), 248-252.

Lamp, J., & Goodwin, C. (1999). Using computer-mediated communications to enhance the teaching of team-based project management. In Hope, B. G., and Yoong, P., (Eds.), *Proceedings of the 10th Australasian Conference on Information Systems* (pp. 484-494). Wellington, New Zealand, 1-3 December.

Leonard-Barton, D. (1988). Implementation characteristics of organizational innovations: Limits and opportunities for management strategies. *Communications Research,* 15(5), 606-31.

Lieberman, A. (1992). The meaning of scholarly activity and the building of community. *Educational Researcher,* 21(6), 5-12.

Mak, S. (1995). Developing a self-access and self-paced learning aid for teaching statistics. *Proceedings of the First Australian World Wide Web Conference,* Ballina, Australia.

Markus, M.L., & Robey, D. (1988). Information technology and organisational change: Causal structure in theory and research. *Management Science*, 34(5), 583-98.

Mayer, R. E. (1992). Cognition and instruction: Their historic meeting within educational psychology. *Journal of Educational Psychology,* 84(4), 405-412.

Miller, S.M., & Miller, K.L. (1999). Using instruction theory to facilitate communication in Web-based courses. *Educational Technology & Society*, 2(3), 106-114.

Mills-Jones, A. (1999). Active learning in IS education: Choosing effective strategies for teaching large classes in higher education. In Hope, B. G., and Yoong, P., (Eds.), *Proceedings of the 10th Australasian Conference on Information Systems*. Wellington, New Zealand, 1-3 December.

Murphy, E. (1997). *Characteristics of Constructivist Learning and Teaching*. Web document. http://ww.stemnet.nt.ca/~elmurphy/emurphy/cle3.html. Last accessed 8 March 2001.

Naisbitt, J. (1984). *Megatrends*. London: McDonald.

Parry, D., Cockroft, S., Breton, A., Abernethy, D., & Gillies, J. (1999). The development of electronic distance learning course in health informatics. In Hope, B. G., and Yoong, P., (Eds.), *Proceedings of the 10th Australasian Conference on Information Systems* (pp. 714-724). Wellington, New Zealand, 1-3 December.

Parsons, T., & Bales, R.F. (1955). *Family, Socialisation and Interaction Process*. Illinios: Free Press.

Pfeffer, J. (1981). Management as symbolic action. *Research in Organisational Behaviour*, 3, 1-52.

Prawat, R. S. (1993). The value of ideas: Problems versus possibilities in learning. *Educational Researcher,* 22(6), 5-16.

Ramsden, P. (1992). *Learning to Teach in Higher Education*. London: Routledge.

Ramsden, P., Margetson, D., Martin, E., & Clarke, J. (1995). *Recognizing and Rewarding Good Teaching in Australian Higher Education*. Canberra: Australian Government Publishing Service.

Resnick, L. B. (1987). Learning in school and out. *Educational Researcher,* 16(9), 13-20.

Riddle, M.D., Nott, M.W., & Pearce, J.M. (1995). The WWW: Opportunities for an integrated approach to teaching and research in science. *Proceedings of the First Australian World Wide Web Conference*, Ballina, NSW, Australia.

Riding, R., & Cheema, I. (1991). Cognitive styles: An overview and integration. *Educational Psychology*, 11(3 & 4), 193-215.

Shipman, F.M., Furuta, R., Brenner, D., Chung, C., & Hsieh, H. (1998). Using paths in the classroom: Experiences and adaptations. *Hypertext 98*, Pittsburgh, PA, USA.

Sproull, L.S., & Goodman, P.S. (1990). Technology and organisations: Integration and opportunities. In Goodman, P. S., Sproull, L. S., et al., (Eds.), *Technology and Organizations* (pp. 254-65). San Francisco: Jossey-Bass.

Staehr, L. (1993). Debating: Its use in teaching social aspects of computing. *SIGCSE Bulletin*, 25(4), 46-49.

Sykes, J.B. (1982). *The Concise Oxford Dictionary of Current English* (7th Ed.). Oxford: Clarendon Press.

Vargo, J., & Cragg, P., (1999). Use of WWW Technologies in IS education. In Hope, B. G., & Yoong, P., (Eds.), *Proceedings of the 10th Australasian Conference on Information Systems*, (pp. 1095-1104), Wellington, New Zealand, 1-3 December.

Willis, J. (1995). Recursive, reflective instructional design model based on constructivist-interpretivist theory. *Educational Technology,* 35(6), 5-23.

Chapter VIII

Delivering Course Material via the Web: An Introduction

Karen S. Nantz and Terry D. Lundgren
Eastern Illinois University, USA

ABSTRACT

The trend toward the use of the Internet in academia is clear with the number of courses using the web for the delivery of course materials ever increasing. There are many positive reasons for using the web in a course, but this raises a number of new issues for faculty, including the time needed and salary concerns. We present a table of six levels of web site use and discuss the major problems associated with the creation and development of web courses. Practical suggestions are offered for dealing with problems from the authors' experiences and the research literature. The final section covers global enrollment, the electronic university, and the trend toward use of the web for the delivery of course materials.

INTRODUCTION

"Education over the Internet is going to be so big it is going to make e-mail usage look like a rounding error."

John Chambers, Cisco Systems
New York Times,
November 17, 1990

Copyright © 2002, Idea Group Publishing.

There are currently over 500,000 courses now available in some form on the Internet (Stevenson, 2000; Telecampus, 2000; Yahoo, 2000), and the media suggest that there will be 2.2 million students taking at least one college course over the Internet by 2002 (Thornton, 1999). Table 1 lists the top 10 eSchools (Stevenson, 2000). The number of universities offering web-based classes is expected to double from 1,500 in 1999 to more than 3,300 by 2004 (E-Learning,

Table 1: The Top 10 eSchools

School	URL (web address)	Notes
California Virtual Campus	www.cvc.edu	California coalition offers over 2,000 online courses.
Indiana University	www.Indiana.edu/~iude	General studies program for multidisciplinary degree.
Penn State	www.outreach.psu.edu	All online students are assigned an advisor.
Rochester Institute of Technology	distancelearning.rit.edu	In top five of U.S. online degree providers.
University of California at Berkeley	learn.berkeley.edu	Excellent academic reputation with wide range of courses.
University of Maryland University College	www.umuc.edu/distance	Has 14 online bachelor's and 10 master's degree programs.
University of Pennsylvania	www.upenn.edu	Wide range of courses from liberal arts through medicine.
University of Washington	www.outreach.washington.edu/dl	Innovative course design and good support for online students.
University of Wisconsin	www1.uwex.edu	More than 150 online courses including business courses.
Western Governors University	www.wgu.edu	Degree credit also available for life experience.

2001). A web-based course can be narrowly defined as one in which little or no instruction takes place in the traditional physical classroom (Navarro, 2000).

Though there is a trend, some say almost a frenzied trend, toward the development of web course materials at colleges and universities, there may be a certain reticence among some faculty and students to "buy in" to the idea that courses delivered on the Internet can replace traditional instructor-delivered courses. First, faculty may think they can do it better by having face-to-face contact with their students and providing a "hands-on" approach to teaching. Second, students may consider Internet technology important but not necessarily a significant part of the educational experience. Internet or computer-based training may be viewed as "busy work." Complicating the picture, there is some pressure from the administration to create web-based courses to generate additional revenue for the institution.

While the move toward Internet-based courses is certainly significant and deserves careful consideration, faculty are far more likely to start by incorporating Internet components into a traditional course; that is, an integration of cyberlearning technologies in the existing curricula (Navarro, 2000). These web-enhanced courses might be considered the transition phase to the new paradigm of Internet-based courses. We define a web-enhanced course as a traditional class with Internet augmentation. Often with no special university sanction or support, faculty members are moving toward the use of the Internet in their courses (Kubala, 1998). The web-enhanced course seems less innocuous. Faculty members still maintain control of the content, and web-enhanced courses are not a "cash cow" for the university.

How prevalent is the use of these web-enhanced traditional courses? The Institute for Higher Education (1999) noted that in 1998, one-third of all classes used Internet resources as part of the syllabus, compared with 25% in 1997 and 15% in 1996. Clearly, using the Internet as a course resource is an important trend for academics.

In many ways, the use of the Internet in academia is becoming a necessity. Research using the Internet brings a myriad of resources that are not available through traditional library print materials from the less mainstream publications to electronic journals. In certain disciplines, such as business and information systems, much current information is largely available only through the Internet. But there is a big difference in the faculty members' perspectives between using the Internet as a research tool and using the Internet for delivery of course content. Using a browser on the Internet hardly prepares faculty for developing a web page. An appropriate analogy would be that reading a book is very different from writing a book.

In an online learning environment, the course materials are of paramount importance. There is no physical faculty presence to motivate and rally the students. Course materials must replace the physical classroom (Twigg, 2001). Interacting with a computer must be sufficiently motivating to keep students' interest.

This chapter discusses the elements of delivering course materials using the Internet. We hope to provide a conceptual underpinning for the use of the Internet that will help remove typical barriers and lend understanding to the process of using a web site for delivering course materials. It is designed for all faculty, from those who have used a web site for years to those who are just considering the use of the Internet to deliver course materials. From the perspective of both research and experience, we present the opportunities and challenges presented by the use of a web site in courses. Some of the topics in this chapter have developed as a result of anecdotal experiences that have never appeared in the literature. These elements should prove useful to anyone interested in the use of technology in the classroom, both academically and administratively.

BACKGROUND

Internet-based courses, also referred to as web-based courses (Mesher, 1999), are defined as those where the entire course is taken on the Internet. In some courses, there may be an initial meeting for orientation. Proctored exams may also be given, either from the source of the web-based course or off-site at a testing facility. The Internet-based course becomes a virtual classroom with a syllabus, course materials, chat space, discussion list, and e-mail services (Resmer, 1999). Navarro (2000) provides a further definition: a fully interactive, multimedia approach.

This is contrasted with a traditional class that meets on a predetermined, regular basis with faculty lectures as the primary method of content delivery. With the availability of web resources in classrooms and computer labs, the traditional class may be enhanced by the introduction of web resources. The web-enhanced course is a blend with the components of the traditional class while making some course materials available on a web site, such as course syllabi, assignments, data files, and test reviews. Additional elements of a web-enhanced course can include on-line testing, a course listserver, instructor-student e-mail, and other activities on the Internet. Navarro (2000) defines this type of course as a "digitized text" approach.

In university courses, faculty have been putting course materials on the Internet (Kahn, 1997), posting assignments, using e-mail for receiving student assignments, adding a discussion web (Hayen, Holmes, & Cappel, 1999), and providing course content with video streaming. The World Lecture Hall (2000) web site contains links to pages created by faculty members worldwide who are using the web to deliver course materials. An Internet search on common course titles will yield many individual course sites. It can be concluded that there are, and will continue to be, many faculty who are adding Internet components to their courses.

THE MAJOR ISSUES

Clearly, there are many positive reasons for using a web site in a course, including greater efficiency in the delivery of materials, providing up-to-the-minute content, enhanced status for the course and faculty, and of course the seemingly inevitable trend to use more technology in education.

But putting a syllabus on a web site for students to access is no simple matter for many faculty (Cooper, 1999). Faculty fear the "programming" connotation of writing Hypertext Markup Language (HTML) code. Even though many of the popular software suites provide a conversion from text-based to web-based documents, the result may be less than satisfactory. Although creating simple web pages can be done with no coding, creating tables, frames, and using graphics may be beyond the expertise of the beginning academic designer. In addition, most faculty have had no training in web design concepts which are substantially different from paper-based design elements. Once acceptable web pages are produced, publishing them to the web, checking hyperlinks, and updating and maintaining the site can be daunting tasks involving the setup and organization of an appropriate server location (web site) and then using the correct file transfer protocol to post the materials to that site.

Another issue is web "glitz." Many students are used to seeing and expect high-impact web pages. But faculty may not have the time nor the expertise to create highly interactive pages with animation, audio, or video streaming. Faculty may feel embarrassed by posting a syllabus that has been simply converted to web (htm, html) format. They can distribute the syllabus in paper format with less disruption of their traditional routines and no pressure to compete with the Internet media.

An additional issue is Internet accessibility. Many business analysts have begun to talk about the "haves" and the "have nots"—those who own a computer and those who do not. When required course content is only available through the Internet, faculty are relying on technology over which they have little control. The proverbial "the dog ate my homework" becomes "the server is down," "my modem died," and "I kept getting a busy signal." This issue of the "have nots" can be addressed with adequate university computer lab open at times to accommodate class and work schedules, but labs traditionally are full to capacity during peak semester times. An answer may be Internet cyber cafes where access time is charged per minute (Gellner, 2000). In the United Kingdom, Jet petrol stations, a subsidiary of Conoco, are providing Internet kiosks (The Western Mail, 2000). Using a smart card system, users access Internet resources ("Log on as you fill up at the petrol station").

Finally, there is the time and pay issue. Faculty are already hard pressed to find the hours needed to keep content relevant. Navarro (2000) suggests that "cyberprofs" typically spend twice as much time developing and teaching web-based courses for the same or less pay than professors teaching traditional courses. The authors' experiences suggest that placing all course materials on a web site, testing, maintaining, and updating them approximately doubles the course preparation time. Even the most minimal use of a web site for delivering course materials is likely to increase course preparation time.

Course Web Site Use

To effectively deal with the above issues, we begin with a classification of the ways that the web can be used in a traditional classroom where lecture has been the primary method of delivery with a defined classroom and meetings. We define six different levels of "web" classes. At the top levels are the Internet-based classes, i.e., the course was created and organized to be web delivered. The middle levels involve a web class that uses the Internet for delivery of content and communication among the course registrants, but also uses face-to-face meetings for some classes, orientation, and testing. At the lowest level, some course materials are simply presented in a hypertext format that replace traditional printed handouts. Table 2 shows the classification levels of academic web pages by typical content and maintenance levels.

Problems

In a 1999 research study, a survey was made of faculty members at a midwestern state-supported university to determine how the Internet was being incorporated into courses (Lundgren, Garrett, & Lundgren, 1999-2000). The results showed that 27.3% of the faculty members think they use the Internet for the delivery of course materials, but we could only verify that 15.6% actually did so. Of this group, the major use was simply the substitution of a web page for the printed page, e.g., the syllabus, schedule, and assignments. Most faculty members (73.8%) updated their sites so infrequently that it must be

Table 2: Classification of Academic Web Pages

Level	Description	Typical Content	Maintenance Level Required	Representative Site(s)
1	Traditional Course Presentation, Basic–Level Course Materials on Web —Internal Links	Instructor data (name, phone, office hours, email address); course materials (syllabus, generic schedule, assignments); noninteractive	Low—static pages after initial upload. Low-volume email correspondence.	www.ux1.eiu.edu/cfiwb www.ux1.eiu.edu/ ~cfjsm/vnhsyl.htm
2	Traditional Course Presentation—Intermediate-Level Course Materials on Web—External Links	All Level 1, some external links, such as textbook and reference sites; non-interactive.	Low—mostly static pages with occasional updates and check-ing of external links. Low-volume email correspondence.	www.ux1.eiu.edu/~cfrc/ index.htm www.ux1.eiu.edu/ ~cseac
3	Traditional Enhanced Course Presentation —Intermediate-Level Course Materials on Web & Web Content Delivery	All Level 2, all traditional course materials posted. Web access in class used for delivery of some course content. Some assignments/ requirements involve interaction, e.g., email submissions, listserve postings.	Weekly updates to schedule, FAQ, course materials, notes to students. Medium-volume email correspondence.	www.ux1.eiu.edu/ ~cfmb/2525.htm www.eiu.edu/~pilotl
4	Traditional Enhanced Course Presentation —Complete Web Content & Materials	All Level 3, course presentations and lectures dynamically available on Web. Data files, links, programs on Web for students. Forms for student "reply" assignments, course evaluations, etc. Link to course grades.	2-3 times per week. Regular updating of grades. Medium-volume email correspondence.	www.eiu.edu/~cis1950 www.advant.net/eiu
5	Web-Delivered Course with Orientation and Testing Meetings	All Level 4 plus any additional materials to allow for full web delivery of course, including audio and video augmentation; multimedia CDs. Few or no regular classes—orientation meeting may be necessary. Testing may be proctored off-site or unproctored on the Web.	Daily maintenance and access by instructor. High level of email correspondence. Regular updating of grades and course materials.	For a listing of hundreds of online courses, see www.ivc.Illinois.edu. www.eiu.edu/~csmith/ eiu4100/eiu4100top.htm webclass.Lakeland. cc.il.us/crincker.
6	Virtual Class	All Level 5 plus on-line testing and orientation. Discussion, chat groups, list serve, email, and other interactive tools;. teleconferencing. No class meetings.	Substantial daily maintenance (average 1-3 hours) by instruc-tor including all course aspects. High level of email correspon-dence.	Most of these courses require passwords or special access information obtained upon enrollment and payment of fees.

concluded that the sites only serve to replicate printed handouts. In a follow-up study at the same university, the number of faculty who used web pages to enhance their courses showed a decrease from the previous year. (Garrett, Lundgren, & Nantz, 2000).

These research results are difficult to understand. Even though virtually all faculty have been provided the appropriate hardware, software, and have an Internet connection in their offices, we found that 22% of the faculty are not ever planning to use a web site for delivery of any portion of their courses. When we accessed the web sites used for classes, we found that most of the faculty members who claim to have a web site are significantly overstating their case. We concluded that there are only a few faculty, less than 5%, who are truly incorporating web technology into their courses in a meaningful way.

There are many reasons why faculty do not embrace new technologies–from a fear of the technology to a lack of reasonable access to the technology. (Rao & Rao, 1999). In our research, we found that level of computer experience is a significant variable why faculty don't have a web site for their courses. Obviously some level of computer literacy is needed for even the simplest web site, and while the university has provided the necessary hardware and software, this does not mean that the faculty have automatically acquired the necessary background for using the Internet in the delivery of their course materials.

One university response to help ensure the success of a technology-enabled course is to provide training for faculty. Unfortunately, this approach does not appear to be particularly productive. Faculty dutifully troop off for the training provided in an environment guaranteed to remind them of their school days as they are instructed to "stay with the group" and to follow the steps in the materials. It would be difficult to think of a less effective way to teach university faculty how to develop a web-enhanced course. Faculty who are trained in the use and integration of new technologies into their courses are not rushing to implement their new-found expertise (Rups, 1999). Anecdotal evidence overwhelming shows that most faculty learn to develop web materials on their own.

University faculty have long demonstrated that they can learn very well on their own if they are provided the materials and incentive. Indeed, to remain contemporary in their fields, university faculty must learn on their own since no one else has the expertise or experience to be their teachers. We suggest that many university faculty may learn new skills and concepts independently and not in a formal classroom situation. From the research cited above, we can say with certainty that virtually none of the faculty received any formal training in web site construction and development.

Some faculty may have access to instructional design experts who can create the sites and add in the "glitz." Our experience shows that the primary responsibility for day-to-day maintenance still falls on the faculty member.

Even with the available technology, training, and resource materials, many faculty still fear the technology. This situation may be exacerbated if there are students in the class who are perceived as more technically savvy than the faculty member. Even if students are not more knowledgeable, fear that they are may be very inhibiting as well as simply fear of showing their ignorance or inability to deal with a technological problem in front of the class.

SOLUTIONS AND RECOMMENDATIONS

We believe that there are compelling reasons to want to use the Internet, and we have devoted a significant amount of time and personal resources to incorporate the technology as fully as possible. For faculty who are interested in moving to the use of the Internet in their courses, this section provides a framework for dealing with the major issues.

Moving Your Course to the Web

The six levels presented above indicate progression from the most basic web-enhanced course to a course delivered fully on the Internet. Faculty would likely proceed through the levels to reach Level 4 for traditional classes unless limited by resources, expertise, and administrative factors. Levels 5 and 6 require significant changes in the academic structure and considerable support of the academic computing environment. Table 3 on the next page summarizes the resources that would be involved in the process of moving your course to the web following the categories outlined.

Although Table 3 shows a summary of the typical resources faculty need to develop web course materials at varying levels, there are other elements that will be just as important in achieving a specific level of web course expertise. The following limiting factors are defined along with suggestions for dealing with them from a practical perspective.

Lack of Experience/Training/Resources

Our experience is that universities typically provide training in the basics. Your university may have licensed a professional commercial package such as Web-CT (2001), Mallard (2001), or BlackBoard (2001) that allows non-technical faculty to create and maintain web-enhanced courses. One of the best ways to learn is to use traditional application packages, such as MS Word, and convert the text to html which can then be viewed by a browser. Beginning web faculty can then see how files are

Table 3: Resources Involved in Moving to the Web

Level	Description	Resources Needed
1	Traditional Course Presentation, Basic-Level Course Materials on Web—Internal Links	Basic computer literacy, web browsing experience. Course site can be created by the faculty member, professional designers, by use of web course applications such as McGraw-Hill's Pageout (2001) or WebCT (2001).
2	Traditional Course Presentation—Interme-diate-Level Course Materials on Web—External Links	Experience with preceding level. Web application packages can be extended or with additional training, a general web development package like MS FrontPage or DreamWeaver can be used.
3	Traditional Enhanced Course Presentation—Intermediate-Level Course Materials on Web and Web Content Delivery	Experience with preceding level. Commit-ment to regular maintenance. Knowledge of e-mail attachments, listserve mainte-nance, or other interactive web applica-tion. Both web application and general web development packages can be ex-tended for this level.
4	Traditional Enhanced Course Presentation —Complete Web Con-tent and Materials	Experience with preceding level. Profes-sional web applications may not be able to accommodate this level without consider-able difficulty. Usually requires consider-able expertise with general web develop-ment packages and some knowledge of HTML and programming concepts.
5	Web-Delivered Course with Orientation and Testing Meetings	All of the above. No additional faculty resources required; academic structural change to allow for registration and other student activities online.
6	Virtual Class	Use of a sophisticated commercial web course package that allows for secure online testing; considerable administrative support and faculty expertise in the selected package.

converted and learn what formatting does not easily convert to html. There will almost certainly be someone who is authorized to help set up a faculty course web site and to provide the basic instructions for posting an html page to a course site.

To develop a Level 4 web site, faculty almost certainly need some understanding of basic programming concepts and HTML. There are many resources on the web that explain elementary html. These guides are a useful resource, as are the many books published on web publishing. These range from the *Internet for Dummies* types to sophisticated books on JavaScript and XML. Elementary html code is simple to understand. It harkens back to the early days of word processing when control codes were inserted into text to indicate formatting. The problem is that html code is cumbersome, and most faculty who really want to use the Internet aren't satisfied with simple html code. Many of the advanced features of web page design require knowledge of common gateway interface (CGI) scripts, Java, or XML (Extensible Markup Language). These require understanding programming constructs and syntax. A valuable resource is any web page itself. The html code can be accessed by viewing the source code in the browser. The code may be specific to certain software tools, but it can serve as a starting point. Cutting and pasting code can get a novice started. Be careful though about any copyrighted pages.

Many universities have instructional support departments that include instructional design specialists. These professionals are invaluable in helping to create course templates—models for the course web pages. Some institutions are licensing professional web software packages with support services to aid faculty in the development of web-enhanced courses. Specific examples are discussed in the next section.

Resources may be the biggest issue here. Navarro (2000) created a set of 4 CDs with 25 lectures in introductory micro- and macroeconomics. The cost of production was $115,000. Each lecture required 24 hours for writing, recording, and editing from the instructor and an additional 162 hours of multimedia development from an assistant. Those time and cost requirements would be daunting to the most ardent cyber learning proponent. Marchese (1998) notes that development costs can be as low as $12,000 per credit hour up to $90,000 per credit hour.

Lack of Access to Technology

We don't see access to technology as an issue, although it is cited in the literature as a reason why academics don't use web technology. Any faculty member who wants to incorporate web technology into classes needs ready access. Since most universities are pushing for web-delivered courses, administrators should be willing to provide the technology. We suspect that many academics who say technology is not available are simply using this as an excuse to avoid the development of web sites for other reasons.

If development tools are used, those technologies must be factored in. One possibility is an on-line campus integrator, such as FirstClass or eCollege.com, which can serve as a web host, provide technical support, and help with course development. Again, costs are an issue with a range of $60-$120 per student per course. Another possibility are web platform providers such as BlackBoard (2001), WebCT (2001), Mallard (2001), and Webmentor. These products are development tools which allow creators to maintain control over the content. While these platforms are designed to be licensed to an educational institution, some such as Blackboard allow faculty to set up a web-enhanced course for free or a modest cost ($100).

Lack of Understanding of Content Presentation

The area of content is one that is understood by many academics. Our earliest attempts at web page design were pretty mediocre because we tried to present web content in the same way that we presented printed content. It doesn't work. For example, in one of our classes, the printed syllabus is eight single-spaced pages. In printed format it works because pages are numbered, headings make organization clear, and students can thumb through pages to reach needed content. Simply converting this to html doesn't work. The eight pages became a master page with links to six other web pages, a course schedule, links to email, and so on. It is a matter of rethinking the access and the purpose of the content.

For the course content, it doesn't work to simply post Powerpoint slides, although they are a tool for organizing and reviewing information. If slides are posted, then make it reasonably simple for students to download the slides (not just view them). Encourage the students to bring the slides to class to use as a format and guide for taking lecture notes. Be prepared for a drop in attendance if you post all of your lecture slides; no matter how often you say it isn't true, some students will see the slides as a substitute for attending class. Finally, all experts would agree that providing pages of text is the death of web-delivered content. Students won't scroll through pages of text or read 10 pages of a case online. The general rule is to keep each web page to about one or two screen pages which translates to a single printed page, and use links to provide navigation. Nantz and Lundgren (1998) provide guidelines for using presentations (slides) in traditional lecture courses.

One of the reasons that web courses don't work is that many of us lived through the old programmed learning paradigm—read a paragraph, answer a question. This methodology is boring and not compatible with today's Web content. Students are familiar with commercial glitzy sites. An academic course site that simply creates text on a web page defeats the purpose of using web pages—of having the ability to create links to interesting sites, to provide graphics, to provide sound and video. For example, consider a course that teaches introductory computer concepts. You are creating a lesson on how a computer works and trying to explain the role of random access memory and the CPU. You could explain in pages of text with even some colorful graphics. But wouldn't it be more interesting to dynamically show what happens to a character from the time the keyboard is pressed until it arrives in RAM, is processed, and stored? However, be aware that the tradeoff for being interesting is that the content might take you hours to create.

Another issue to consider is that on the web we have to teach without lecturing. Simply putting lecture notes on the web will create uninteresting content. There is no way to adjust the content based on our students' reactions since we may never see our students.

Lack of Confidence in the Stability of Technology

One of the issues for us has been the stability of technology. Over 90% of instructors report frequent problems (Navarro, 2000). We are blessed to be in a relatively new building with computer technology, Internet access, and projection capabilities available in each classroom. There's nothing worse than preparing an interactive lecture with dynamic linking only to have the server down during class. There is no way to duplicate the effect with transparency backups. One of us once spent nine hours putting together a dynamic lecture using web resources that couldn't be delivered because the university servers were down. If you are relying on the web to present vital course materials, then you need to be flexible. Students have to have access if the web is their only recourse. Plan for technical problems and alert students to contingency plans. Common complaints include web congestion, inactive links, email formats that don't accept or send attachments, and incompatible Internet browsers.

Lack of Time and/or Incentives to Incorporate Technology

Many of us are in an environment where research is valued more than teaching, although good teaching is considered necessary. Innovative teaching is often praised, but not rewarded. For example, in our yearly evaluations, refereed journal articles are absolutely essential to advancement. Creating web content for courses is far, far down the list.

There is no doubt that creating web pages is going to take significant time. We estimate that putting a course online is going to take at least twice the time it would take to put a traditional lecture course together. If you are proposing web-enhanced courses, make sure that you build in sufficient development and maintenance time. In Table 2, we have tried to suggest the maintenance time needed. In addition, every time the course is offered, content must be "tweaked" and links checked. The one constant on the Internet is change.

Navarro (2000) also notes that "cyberprofs" are reporting strong negative reactions from their colleagues. This reaction may occur because colleagues don't understand how a web-enhanced course works. If you are sitting in your office answering student email or creating course content, you don't appear to be teaching. At our university, anecdotal evidence suggests that colleagues have been unfairly treated by their peers for introducing significant web development in classes. The amount of work required has not been translated equitably into faculty evaluations.

One incentive that has been used at many universities is the use of royalties. Professors are paid for their web-based and web-enhanced course content. Again, this can create animosity with colleagues who may not have the opportunity to receive additional compensation for course content. Arthur Levine of Columbia University has suggested that we may all need academic agents who will negotiate deals for the rights to popular web courses (Twigg, 2001).

Incentives may become a much larger issue. A company called UNext.com is offering universities the opportunity to use their names and faculty expertise to market courses worldwide. UNext.com was originally known as Knowledge Universe. The contracts make it clear that content is the property of the university and not individual faculty members. Each university then decides faculty compensation. Estimates of revenues for participating, high-recognition universities could be as high as $20 million over a five to eight year period (Twigg, 2001).

Student Attendance

Attendance in a web-enhanced course will drop. If a faculty member replicates course presentations (outlines or Powerpoint slides), student may perceive that attendance is not necessary. Research over the past 20 years shows that the average college attendance rate is about 70% and the rate is remarkably consistent across university populations. However, the distribution across classes at any given university is quite skewed. Courses that have sustained attendance rates over 95% are extremely rare with most classes in the range of 50 to 90% (Lundgren & Lundgren, 1996).

If your class has an average attendance rate of 70%, then adding a Level 3 web-enhanced site will drop that rate. We do not have population data, but a drop of 20% would be predicted and you will find that some classes in the last third of the term will have very low attendance as the students come to perceive that the web materials do indeed substitute for some class attendance. "Getting the notes" from a fellow student has been replaced by the course web site which is easier and more accurate. Of course you can force attendance in a number of ways, most of which are perceived by the students as punishment for lack of attendance. The student must perceive an added value for class attendance. This should be a positive enhancement of learning rather than simply penalties by quizzes or loss of course points for not attending. Frankly, we have no feasible answer for this problem and believe that perhaps we should accept that student learning with a web-enhanced course does not require attendance levels comparable to the traditional course.

Intellectual Property Issues

Many faculty do not consider the issue of copyright and intellectual property when course materials are developed (Rueter, 2001). The issue is not clear cut. Most faculty believe they own their own course materials. This is often not the case. The issue becomes even muddier when entire courses are delivered on the web (Levels 5 and 6). A course that you developed could be offered by the university with someone else teaching it, without your consent or knowledge. Earnings from distance learning are viewed quite differently by faculty and administration (Guernsey & Young, 2001).

Dennis Thompson, Associate Provost and Professor of Government at Harvard University, notes that intellectual property issues associated with information technology are the most contentious issues on campuses. The current laws do not adequately address the problems. Thompson (1999) suggests that the following are intellectual property issues for academic web sites.

- Should university administrators have control of web-delivered course content and gain the potential profit for the university?
- Are web sites copyrightable? Do they fall under patentable products?
- Are interactive web pages different from other pedagogical outputs?
- If the university has provided substantial resources for the production of academic web pages, does that imply ownership in some form?
- Are web pages "work for hire" which would be a deviation from traditional exemptions for faculty?
- Is a non-required academic web page different from required syllabi for courses?

Thompson notes that although law and current practice are not clear cut "when the university decides at the margin to expend funds to provide a faculty member with an exclusive resource instead of allocating the same funds to a collective resource, it has a legitimate claim on the products created using the exclusive resource."

Protection is a two-way street. On the one hand, distance learning provides universities a "virtual university." Faculty teaching resources become digital and may or may not be taught by the author. On the other hand, faculty could become independent contractors and offer the web courses at multiple universities. The whole idea of ownership of courseware presents new challenges. Thompson (1999) also notes that universities should not claim Internet courseware for three reasons: 1) There are no distinguishable differences between web-delivered and traditional-delivered courses. The university is not entitled to ownership. 2) University ownership of Internet courseware creates inequity problems and, in fact, penalizes faculty for being innovative. 3) Treating Internet courseware differently encourages faculty to look for other distribution methods.

The American Association of University Professors (2001) believes that course materials, digital or not, should be considered a professor's property. This view is also supported by the National Education Association. John Reuter (2000) notes that faculty can take specific actions: 1) claim copyright and intellectual property rights to any academic materials they create; 2) make clear any differences between your intellectual property and materials created with university supports; 3) give up royalties to gain intellectual property control; and 4) unless clear ownership is agreed upon, move resources to off-campus servers.

Many faculty contracts are addressing the intellectual property issue. The University of Hawaii contract says, "That in most cases professors own exclusive rights to their work, and that the university cannot reuse an on-line course without the professor's permission." What faculty fear most is the "Player Piano Scenario" where courses are created and then played and replayed without faculty approval.

Lack of Administrative Understanding of Quality Teaching

There is a popular administrative myth that "quality" classroom time involves meaningful discussion, question-and-answer discourse, and significant teacher-student interaction. The implication for the web-based course is that there must be a web substitute for the discussion, discourse, and interaction using web discussion, chat groups, interactive sessions, and lots of e-mail. Unfortunately, the myth is false. Visiting virtually any traditional class in operation today will quickly dispel the myth. Watch the teacher struggle to get students to say anything in class. Notice that the majority of students have no interaction with the teacher and have contributed absolutely nothing in the classroom. Many of the students clearly prefer this type of course where they sit quietly taking notes and are not required to ever respond in any way.

If faculty develop a web-enhanced course following the myth of preserving student interaction, they will be quickly mired in web activities that consume the majority of their time with no observable educational payoff. Discussion groups on the web often quickly degenerate as students write "from the heart" as opposed to critical analysis. E-mail lines between the student and teacher become a conduit for students to ask about their grade, their last assignment, why their grade is not as expected, how to calculate their grade, etc. If you decide to include interactive components on your course site, be aware that these elements can consume the majority of your time and may leave you with a unfavorable experience.

The less subtle problems that stem from a lack of administrative understanding include difficulties in obtaining resources, especially release time for the initial development of a web class, lack of understanding about how many hours are needed to run the course when you aren't standing in front of a classroom, email overhead, managing listservs, and problems in obtaining reasonable hardware and software to develop and maintain a web course.

Our best advice here is to recognize that this may be the default condition and to be pleasantly surprised when you receive administrative support. We caution faculty to proceed slowly when developing a web course and to be very, very realistic in what you hope to accomplish. Though this may sound overly pessimistic, we are optimistic that this situation will be increasingly ameliorated as administration becomes more computer and telecommunications literate.

FUTURE TRENDS

Global Enrollment and the Electronic University

Many universities are seeing the potential of a global enrollment community. If courses can be offered on the Internet, there are no geographical barriers to enrollment. This is a positive for students, who can choose to take a course from any university offering it. Indeed, electronic degrees are being offered by many universities. Many universities will have difficulty capitalizing on web-based course potential. First, they are not in the product marketing business with the appropriate venture capital to support high-end Internet learning. Second, institutions don't have the creative development staff to support such ventures. Third, universities are slow to react to market changes. Think about how long it takes to get curriculum changes through the system.

An issue for the electronic university is what value is placed on the social and interpersonal interactions between students and faculty in a classroom. On the one hand, many students come to class to simply write down every word on the transparencies and occupy the seat. But, students need the social interaction that comes from diverse student populations. If we are awarding degrees for interacting with a computer terminal, what kind of potential employees are we preparing? Students need to be able to interact socially and understand the differences between personal interactions (roommates, friends, etc.) and business interactions. Most freshman are socially inept. It takes the maturity of four years of different social interactions to prepare a student for a career. A university education is not about the isolation of solitary learning.

We believe that students don't attend college just to get canned courses. Students want interaction with faculty members who can mentor them. All of us have had experiences with students we have significantly impacted in some way. Would that same impact have occurred if you had never seen the student face to face?

As our research has shown, there is no faculty rush to embrace the Internet in the delivery of course materials. Further, there will be no rush in the near future in spite of administrative attempts to encourage Internet use. Faculty will continue to say that they are interested in using the Internet because it is expedient, but this will have little effect on the actual number of faculty who use it. In part, this goes back to the fear of being replaced by a machine. Will faculty be hired on a contract basis to design courses based on a certain expertise?

Another issue for us is the effect on the faculty member. We're in teaching because we enjoy interacting with students. One of the perks is a somewhat flexible schedule. In a cyber environment, a faculty member is freed from traditional lectures but tied to the computer with email and bulletin board traffic. Especially in a global environment, students may expect you to answer email at night and on weekends. Navarro (2000) reports that 29% of cyber instructors enjoyed cyber teaching less than regular teaching.

As younger, more computer-literate faculty emerge, there will be a slow and steady move toward the incorporation of Internet components into university courses. Many educational pundits believe that we are moving into a new learning paradigm with the integration of technology into our schools (Von Holzen, 2000). This new educational model envisions a complete shift in course delivery from the traditional lecture classroom to on-demand, flexible learning through the use of telecommunications technology or "just-in-time" learning. In this paradigm, the faculty will become the designers of interactive course materials. With this new paradigm, many of the issues discusses in this chapter may take care of themselves. In the meantime, faculty who are considering web-enhanced or web-delivered courses need to be aware of the issues. But, most of all, we believe that learning is more than just content delivery; we need to create learning environments whether they are in the classroom or in cyberspace.

REFERENCES

American Association of University Professors. *Report on Distance Learning*. (www.aaup.org/dirptext.htm) as of January 26, 2001.

Blackboard. Commercial academic site for delivering course materials at www.blackboard.com as of June 18, 2001.

Cooper, L. (1999). Anatomy of an online course. *T.H.E. Journal*, 26(7), 49-51.

E-learning spending and enrollment to grow steadily. (2001). *College Planning and Management* 4(2), 10.

Frizler, K. (1999). *Designing Successful Internet Assignments*. Syllabus, 12(6), 52-53.

Garrett, N. A., Lundgren, T., and Nantz, K. S., (2000). Faculty course use of the Internet. *Journal of Computer Information Systems*.

Gellner, E. (2000). *Java's Cyber Espresso Bar*. (www.javas.com/home.asp) as of November 6, 2000.

Guernsey, L. and Young, J. R. (2001). *Professors and Universities Anticipate Disputes Over the Earnings from Distance Learning*. (www.chronicle.com/colloquy/98/ownership/background.shtml) as of January 24, 2001.

Hayen, R. L., Holmes, M. C., and Cappel, J. J. (1999). Enhancing a graduate MIS course using a discussion Web. *Journal of Computer Information Systems*, 39(3), 49-56

Institute for Higher Education (1999). *Distance Learning in Higher Education*, Washington, DC.

Kahn, R. L. (1997). Creation and maintenance of a syllabi Web site: A case study. *A Supplement to T.H.E. Journal: The Internet in Education*, 23-26. www.arsc.sunyit.edu/~com400/res1.html.

Kubala, T. (1998). Addressing student needs: Teaching on the Internet. *T.H.E. Journal*, 25(8), 71-74.

Lundgren, T., Garrett, N. A., and Lundgren, C. (1999-2000). Student attitudes toward Internet course components. *Journal of Computer Information Systems*, 40(2), 64-68.

Lundgren, T. and Lundgren, C. (1996). College student absenteeism. *Proceedings of the 1996 Delta Pi Epsilon National Research Conference*. Little Rock, Arkansas, pp. 71-78.

Mallard (2001). *Asynchronous Learning on the Web homepage* at www.ews.uiuc.edu/Mallard/ as of June 18, 2001.

Marchese, T. (1998). Not-so-distant competitors: How new providers are remaking the postsecondary marketplace. *AAHE Bulletin*, May.

Mesher, D. (1999). *Designing Interactivities for Internet Learning*. Syllabus, 12(7), 16-120.

Mc-Graw Hill's PageOut (2001). www.pageout.net as of 6/11/2001.

Nantz, K. S. and Lundgren, T. (1998). Lecturing with technology. *College Teaching*, 46(2), 53-56.

Navarro, P. (2000). Economics in the classroom. *Journal of Economic Perspectives*, 14(2), 119-132. (www.powerofeconomics.com/I.2-Research-Studies.htm) as of March 1, 2001.

Rao, P. V. and Rao, L. M. (1999). *Strategies that Support Instructional Technology*, Syllabus, 12(7), 22-24.

Resmer, M. (1999). *IMS: Setting the Course for Distributed Learning*. Syllabus, 12(7), 10-14.

Rueter, J. (2001). *Modular Courses and Intellectual Property Rights*. (web.pdx.edu/~rueterj/rlw/modular.htm) as of January 24, 2001.

Rups, P. (1999). Training instructors in new technologies. *T.H.E. Journal*, 26(8), 67-69.

Stevenson, N. (2000). *Distance Learning Online for Dummies*, New York: IDG Books Worldwide (now Hungry Minds).

TeleCampus. (2000). Specializing in Online Learning web site at telecampus.edu, as of April, 2000.

The Western Mail (2000). Log on as you fill up at the petrol station. Cardiff, Wales, p. 25.

Thompson, D. (2001). *Intellectual Property Meets Information Technology*. (www.educause.edu/ir/library/html/erm99022.htm) as of January 24, 2001.

Thornton, C. (1999). Back to school, Web-style. *PC World*, 17(7), 39-40.

Twigg, C. A. (2001). *Who Owns Online Courses and Course Materials? Intellectual Property Policies for a New Learning Environment*. (www.center.rpi.edu/PewSym/mon02.html) as of January 24, 2001.

Von Holzen, R. (2000). *A Look at the Future of Higher Education*. Syllabus 14(4), 54-57, 65.

WebCT. (2001). *The e-learning link web site* at www.webct.com, June 18, 2001.

World Lecture Hall. (2000). *Academic Computing and Instructional Technology Services* (web site at www.utexas.edu/world/lecture) as of April 2000.

Yahoo (www.yahoo.com/education/distance_learning) as of February 2000.

Chapter IX

Bridging the Industry-University Gap: An Action Research Study of a Web-Enabled Course Partnership

Ned Kock
Temple University, USA

Camille Auspitz and Brad King
Day & Zimmermann, Inc., USA

ABSTRACT

This chapter discusses a course partnership involving Day & Zimmermann, Inc. (DZI), a large engineering and professional services company, and Temple University. The course's main goal was to teach students business process redesign concepts and techniques. These concepts and techniques were used to redesign five business processes from DZI's information technology organization. DZI's CIO and a senior manager, who played the role of project manager, championed the course partnership. A Web site with bulletin boards, multimedia components and static content was used to support the partnership. The chapter investigates the use of Web-based collaboration technologies in combination with com-

Copyright © 2002, Idea Group Publishing.

munication behavior norms and face-to-face meetings, and its effect on the success of the partnership.

INTRODUCTION

Industry-university partnerships, particularly those involving research universities, are commonplace and on the rise (Burnham, 1997). They allow industry access to quality research services at subsidized costs, as well as to potential future employees while still in their formative years. Universities benefit from such partnerships through research grants that complement dwindling government funding, and student exposure to current "real-world" problems and issues.

Some sectors of the economy are more active than others in research involving industry-university collaboration. The manufacturing sector is arguably the most active. In 1998, the National Coalition for Advanced Manufacturing, based in Washington, DC, released a report on the topic covering a wide range of industries. The vast majority of the companies surveyed for the report praised the concept and highlighted the crucial importance of industry-university partnerships for competitiveness improvement. One association of manufacturers in particular, Sematech, made up of companies in the U.S. semiconductor industry, stated that a considerable portion of its membership had been literally rescued from their competitiveness downslide by industry-university research partnerships (Wheaton, 1998).

Irrespective of economic sector or industry, the vast majority of industry-university partnerships are of the *research partnership* type, which predominantly involves applied firm-specific research. In this type of partnership, funding from the industry partner is received in exchange for "intellectual horsepower" in the form of research services and technology transfer (Hollingsworth, 1998). In science-based fields, universities focus on basic research, and the main interest of industry partners is in the commercial and industrial implications of a scientific project and how they can be taken advantage of by internal research and development departments. In less science-based fields, the solution of technical problems is a major concern of industry. In all fields, the exchange of knowledge in techno-scientific communities is a crucial element of interaction in research partnerships (Meyer-Krahmer, 1998).

A much less common type of industry-university partnership is what we refer here to as a *course partnership*, which gravitates around a regular university course (or set of courses) rather than a research project or program. In these types of partnerships, the industry partner agrees to sponsor one or more courses in which the students are expected to apply concepts and theory learned in class to the solution of some of the industry partner's key problems. Students benefit from the direct contact with the industry they are likely to join after they graduate as well as professional relationships they are able to establish during the course.

This chapter discusses a *course partnership* involving a large engineering and professional services company, and a public university, both headquartered in Philadelphia. An action research study of the course partnership conducted between May and July of 1999 is used as a basis. The main goal of the course was to teach students business process redesign concepts and techniques, which were used to redesign several real processes at the industry partner. One salient aspect of this action research study is the role played by a Web-based collaboration system as a communication hub and information repository during the course partnership, which is investigated in light of previous empirical research and key theories. Like typical action research studies (Checkland, 1991; Lau, 1997; Peters and Robinson, 1984; Winter, 1989; Wood-Harper, 1985), ours aimed at providing a service to the research clients (Jonsonn, 1991; Rapoport, 1970; Sommer, 1994), while at the same time performing an exploratory investigation of the effect of Web-based collaboration technologies on course partnerships. The research clients in question were the students and the industry partner. Also, in line with a subclass of action research, namely participatory action research (Greenwood et al., 1993; Elden and Chisholm, 1993; McTaggart, 1991; Whyte, 1991), one of the research clients, the industry partner, participated actively in the compilation and analysis of the exploratory research data, as well as in the interpretation of the findings, including the writing of this chapter.

OBSTACLES TO COURSE PARTNERSHIPS

The wide proliferation of research partnerships in the U.S. and several other countries (Cabral, 1998; Jones-Evans, 1999; Saegusa, 1997; Wong, 1999) can be explained by the incentives to those who directly participate in the partnership. The benefits for industry partners, faculty and students involved range from knowledge acquisition to financial incentives. Often, research partnerships reward research and development department members with workload reduction, increased productivity and knowledge acquisition. Faculty and students are rewarded with funds to support their research and exposure to industry-specific problems and issues outside the scope of university education.

Course partnerships, on the other hand, often fail to benefit a key group of players – the faculty developing and teaching the courses. Most course partnerships involve the adaptation of existing university programs or the creation of new programs to address the needs of a particular industry or company (Mengoni, 1998). In these "wholesale partnerships," the industry benefits from a university program better tailored to its needs, and the university as a whole from an increase in enrollments. The faculty who teach those courses, however, are rarely provided with any direct incentive to participate in such partnerships, in spite of the extra work required to develop new or adapt existing courses to the new program.

A possible alternative to overcome the barrier above is for universities to stimulate and provide the necessary infrastructure for faculty to lead the development of course partnerships on a course-by-course basis, which could potentially lead to better aggregate results in terms of tailoring courses to industry needs. This new approach could be implemented by supporting the development and teaching of specific courses in close collaboration with industry partners, who would provide funding to compensate faculty for their participation and cover other expenses such as specific equipment and software needed to implement the partnership. Key potential benefits of course partnerships for students and industry partners are listed in Table 1:

Assuming that the problem of lack of direct incentives for faculty is solved, key obstacles to course partnerships still remain. Some of these stem from difficulties in the communication and coordination between industry and university participants. Industry and universities often have different organizational cultures, languages and values, which pose communication difficulties. Their members follow different work schedules, are rarely co-located, and have different and sometimes conflicting goals (Brannock, 1998), which create coordination difficulties.

Table 1: Potential Benefits of Course Partnerships

Benefits for students	**Benefits for industry partners**
Putting concepts and theories learned in class in practice, which adds a new and valuable "real-world" dimension to the learning process.	Hiring selected students with top potential, and whose behavior and values match the firm's internal culture, customer orientation and mission.
Experiencing first-hand professional issues in their chosen fields.	Creating the appropriate climate for change due to the infusion of new ideas.
Establishing company contacts that may lead to future employment.	Absorbing new concepts and ideas that may be used to boost competitiveness.

In addition to communication and coordination difficulties, a key obstacle to course partnerships is the extra time commitment required from both industry as well as university participants. It can be inferred from careful inspection of Table 1 that the more company members are directly involved in the partnership, the better. For example, the more company members observe students in action during the course partnership, the more accurate will be the identification of future "stellar" employees. However, work pressures may make it difficult to motivate a critical mass of employees to participate actively. Also, faculty teaching such courses must to be willing to take on heavy project management responsibilities in addition to normal teaching duties (Lee, 1998).

THE ROLE OF WEB-BASED COLLABORATION TECHNOLOGIES

The deployment of standard Web-based technologies in organizations opened up new opportunities for the development of Web-based systems to support inter-organizational collaboration. Since course partnerships are, by definition, inter-organizational initiatives involving at least two different entities, i.e., an industry partner and a university, they are prime candidates for the use of Web-based collaboration technologies. The availability of a common infrastructure, the Internet, allows for fast implementation of low-cost Web sites with effective communication support and data centralization features. Such Web sites can potentially be used to overcome several of the obstacles outlined in the previous section in the context of the course partnership discussed in this chapter.

As stated earlier, the course partnership described here involved the establishment of teams whose main goal was to improve several business processes of the industry partner. For this to happen effectively, both industry partner members and students had to agree on the basic concepts, techniques and language used in process improvement initiatives. In this context, Web-based bulletin boards were seen as likely to be useful complements to face-to-face meetings by allowing participants to conduct part of their interaction in an asynchronous and distribute manner, using standard Web browsers. Previous research suggests that the use of electronic bulletin boards is likely to decrease the amount of time required from each individual member of a process improvement team, without any loss of quality, provided that Web-based interaction is used in combination with face-to-face meetings (Kock, 1999; Kock and McQueen, 1998).

Most process improvement initiatives involve the subsequent phases of process selection, modeling, analysis and redesign (Davenport, 1993; Davenport and Short, 1990; Hammer and Champy, 1993; Harrington, 1991). Given that progressing through these phases generates a large amount of documentation, another useful function of a Web site would be that of providing a central repository for the documentation generated by process improvement teams. The documentation of a given team could also be made available to all teams, so opportunities for integration of the outcomes from different teams could be identified and taken advantage of.

Finally, the combined use of a Web site for communication and centralized data storage would likely improve coordination of the work of different process improvement teams and reduce the amount of time and effort required for that coordination. The project manager would be able to monitor the progress of each team vis-à-vis the progress of the other teams through the Web site, without having to rely only on time-consuming face-to-face meetings, and take action when needed. General project instructions and guidelines could be provided once for everyone through a specific area of the Web site, rather than repetitively to each team. The sharing of data among teams would likely enable some teams to avoid mistakes made by other teams, as well as reuse interesting process designs and related ideas.

ACTION RESEARCH STUDY: DAY & ZIMMERMAN, INC. AND TEMPLE UNIVERSITY

This section describes the course partnership and discusses the key role that a Web site has played in the success of the partnership. While this chapter has three co-authors, including the instructor who was one of the investigators, this section was written in most part by industry partner co-investigators. It reflects their perceptions of the partnership, which the instructor validated through participant observation and research data analysis (this is mentioned so the reader can better appreciate the narrative). It is clear from this section that several of the expected technology benefits above were realized, but some were not. The next section summarizes and discusses both types of effects as "lessons learned" from the action research study.

Initial Contacts and Meetings: Do We Really Want to do This?

In late April of 1999, the CIO of Day & Zimmermann, Inc. (DZI) received a letter from the instructor (first author of this chapter) inquiring if DZI was interested in partnering with Temple University for a graduate-level course in business process redesign with applications of Internet collaboration technologies. From the outset, the CIO seemed to believe that this project would benefit DZI's enterprise IT organization (eIT). A meeting was arranged and the instructor went to DZI's headquarters in Philadelphia to meet with the CIO and his management team. At the meeting the instructor presented his ideas about the course partnership and proposed an implementation plan. eIT's management found the idea intriguing. The consensus was that a collaboration of this order presented a number of opportunities as well as a number of obstacles.

eIT had recently gone through a major transformation, which included a Capability Design project. This project involved many eIT employees in various teams that evaluated high-level processes and designed new processes as a foundation for the newly reorganized unit. The proposed partnership with Temple University was seen as likely to enable eIT to leverage some of the momentum created by the Capability Design project. In addition to gaining valuable process design experience, the eIT employees involved in the Temple project would, working with the student team members, be able to identify real process improvements that could be applied once the course was completed. Temple's offer of evaluating and redesigning tactical-level processes was seen as a good complement to the high-level process redesign of the earlier Capability Design project.

The most significant obstacles faced by eIT to the implementation of the course partnership were related to time and resource constraints. In addition, there was some concern about whether the redesigned processes would actually be implemented once the course was over. eIT's management and staff were fully engaged and had little or no spare time in which to fulfill the obligations of this type of project. In response to this concern, the instructor proposed that a Web site be used to support the project as a data repository and communication tool. And since the course was being taught both online and in lectures, and the entire course material was to be posted online (including streamed video-clips of lectures); any eIT employee would be able to take the course online, even if he or she was not able to sit in the face-to-face class meetings. An additional, albeit less tangible benefit was the opportunity for the company to observe and evaluate young talent for potential hire into eIT or DZI. It was determined that the benefits outweighed the drawbacks even when taking into consideration that the redesigned processes may not be implemented. The course partnership was formed.

Selecting Projects and Defining Their Scope: Managing Expectations

Several processes within eIT were identified for the project. The following processes were selected jointly by eIT and the instructor: Asset Management,

Help Desk Call Response, SLA (Service Level Agreement) Development, New Employee Account Set-Up and Key Person University (KPU). A process owner from within eIT was assigned to each of the selected processes. Each process was then assigned a team of students, who were expected to work closely with the process owner. A key element of this project was the working relationship between the process owners and the student teams. Everyone involved was forced to stretch intellectually, organizationally and socially to coordinate multiple schedules and skills. It was evident from the beginning that some groups would perform at a higher level than others. The instructor cautioned eIT management about setting expectations too high. The goal for the university was to teach the students about process redesign and collaborative work methods. eIT had to understand that even though the project very closely resembled a consulting project, the actual redesign results might not be functional or feasible.

eIT strongly believed that this partnership and the project as a whole would have a higher success rate if the focus remained inside eIT and thus limited the amount of involvement required from other DZI businesses and staff units. eIT also assigned one of its senior managers as a full-time project manager for the course partnership. This level of commitment on the part of the sponsoring company, as well as the decision to fund the partnership through a cash grant, clearly signaled the significance of the project for both the process owners and the student teams. Temple University provided two technical support specialists who were available to assist both eIT and the student teams with various technical issues that arose during the course.

The overall scope of the course was clearly defined in the course outline. Three reports were to be developed. Report 1 included a contextualized description, model and list of problems associated with the current process, as well as desired achievements of the redesign. Report 2 included the redesign guidelines used and how they were used, and the redesigned process model. Report 3 consisted of an analysis of three different IT solutions to implement the redesigned process, as well as a cost/benefit analysis and an implementation plan for each solution. Specific requirements for the individual processes varied. Of the five selected processes, all were existing processes within eIT with the exception of the KPU. Therefore, development of the first report for the KPU process team was more complicated because they first had to develop a process that they could then redesign. Some of the processes such as the Help Desk and New Employee Account Set-up required greater involvement from extended team members and eIT customers.

Creating a Web Site: Not Only a Communication Tool

The development of a Web site to support the partnership was a key selling point to eIT from the outset. The instructor created a Web site as the hub of information and collaboration for the course. The Web site contained all the course material, lecture slides, course outline, contact information, pictures and video clips of student teams, discussion threads for each of the processes and a "tools" area that provided detailed tutorials that explained how to create a Web site with similar functionality. It was through this Web site that eIT employees could "attend" the course and all the process team members communicated asynchronously (see Figure 1).

The team members used the discussion threads available on the Web site to post and resolve many issues while working on various draft reports. The process teams met regularly to discuss and review these reports. The Web site enabled all parties involved to access and review draft documents prior to a meeting. Meeting time was therefore maximized and used to discuss modifications to the document content as opposed to being wasted while all present familiarized themselves with a revised document.

Figure 1: Main Page of the Web Site

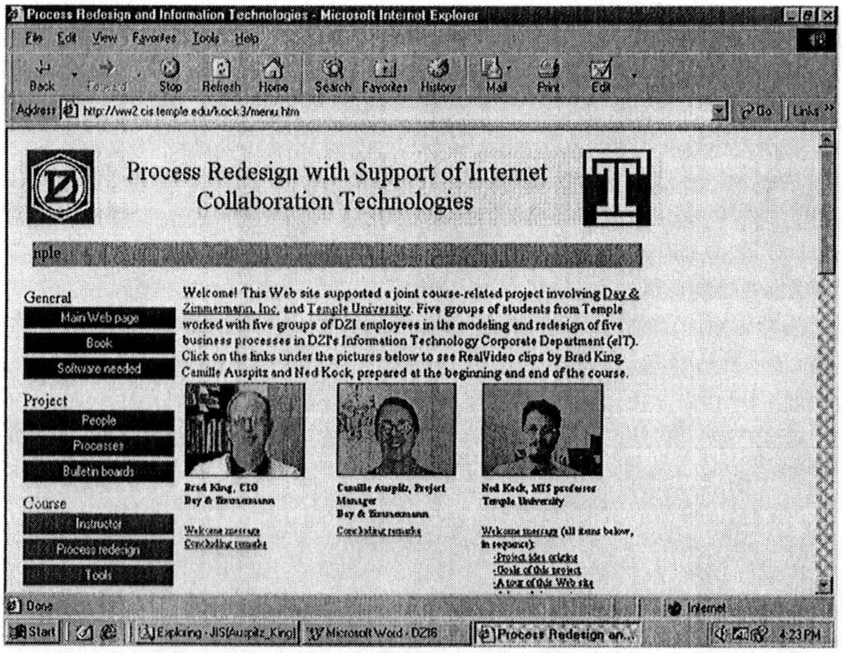

It was essential that the skills needed to take advantage of the Web site were developed quickly, which was facilitated by the fact that all features of the Web site required only a standard Web browser and some free players to be fully utilized. DZI's project manager strongly encouraged all the process owners to access the Web site and review their discussion threads at least twice a day; the instructor did the same for the students. Process owners and student team members were asked to review the progress of the other process teams as well. Each process team developed their own schedule and style to accomplish the requirements of the course. Some teams met weekly; others twice-weekly; others still met only every other week; all with more or less the same level of success in regards to completing the reports.

The Web site was also made available to all eIT employees throughout DZI (over 100 people). While interest and participation from employees not directly involved in the course partnership fell below expected levels, there was enough interest in the technology behind the site to initiate conversations about applying similar technology within the organization. It is essential that diverse IT organizations, such as eIT, share information and collaborate on enterprise issues. Today, well after the course partnership was concluded, eIT has a very active discussion site that addresses numerous topics and issues throughout the organization. While the underlying technology is different, using Lotus Notes and Domino instead of Microsoft FrontPage and Internet Information Server (IIS), the capability is very similar and mirrors that of the Web site developed to support the course partnership.

Electronic Versus Face-to-Face Meetings: Which One to Choose and When?

There were advantages and disadvantages to using Web-based collaboration technologies to address and resolve issues and interact with project team members. The teams were very diverse. Utilizing the online electronic discussions (see Figure 2) was often seen as an effective way to overcome language barriers by formally defining process-specific terms and to clarify issues not properly addressed in face-to-face meetings. In addition, as stated earlier, the ability to maximize precious face-to-face time by enabling all team members to have access to documentation and discussion topics prior to a meeting was seen as essential. In order to accomplish the objectives of the course, the electronic collaboration capability also enabled team members who were not able to attend a meeting to contribute as well.

However, most participants felt that it was necessary to complement the Web site interaction with face-to-face meetings. From a project management perspective, the face-to-face meetings were seen as particularly effective for

Figure 2: General Discussion Board for the Course Partnership (each process team had its specific discussion board)

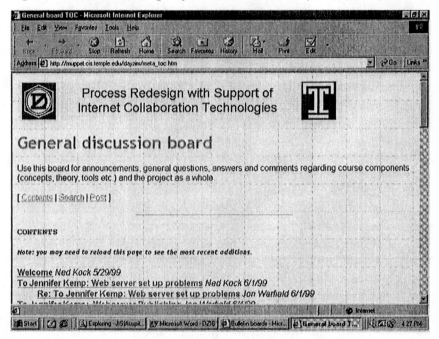

building and ensuring consensus. From a corporate standpoint, it was during the face-to-face meetings that the "stellar performers," both students and process owners, could be identified. The same was true for student evaluation from an academic standpoint. While the reports were collaborative efforts and discussions often involved many team members, it was in the meetings and final presentations that the natural leaders and above-average achievers became evident.

Going the Extra Mile: Project Review, Pizza Party and Final Presentations

Throughout the five weeks of the course, DZI's project manager provided eIT management with regular status reports summarizing the progress of the various teams. This enabled DZI's CIO to remain up to date as the course progressed. However, the CIO was also interested in the technology used to support the project, the general subject of process redesign, the participation of his staff in the project and what they had learned about their specific processes. In order to provide him with a clearer perspective of his staff's experience, a project review was scheduled in which the eIT process owners would offer the CIO and an external consultant from the Concours Group a presentation detailing the teams' progress.

The review took place on June 22, 1999 just three days before final reports and final student team presentations were due. Student team members were asked to be present for the review as well. This allowed each presentation to be followed by a dynamic discussion between the process owner, the student team members, the CIO and the external consultant. These discussions ranged from delving deeper into process descriptions and redesigns to lively forays into technologies' future. A fair amount of time was spent discussing how technology affects business processes and the growing role technology will continue to play in strategic business planning initiatives.

DZI's eIT organization was very satisfied with the progress and outcomes of this collaborative project and was able to demonstrate that enthusiasm by hosting a party following the project review presentations. All the students, process owners, extended team members and eIT management sponsors enjoyed a casual party at a local pizza restaurant. This setting was conducive to informal and lively conversations and made the CIO, the instructor and the external consultant available to many of the project participants. Given that the course had been very intense for all the participants, the gathering also allowed for sharing of personal stories and interests, and seemed to have created a relaxed atmosphere and an engaging and satisfying note on which to end the project.

The final presentations took place on the last day of the course and were held in DZI's boardroom. Each of the student teams prepared a PowerPoint presentation that summarized Reports 1 and 2 and focused on the content of Report 3, which, as mentioned before, consisted of an analysis of IT implementation solutions for the redesigned process, as well as a cost/benefit analysis and implementation plan for each IT implementation solution.

In many of the final presentations the IT solutions in conjunction with the redesigned process were viable when considered independent from one another. What came to light essentially during these presentations was the fact that since all the processes were within the domain of eIT, they were all in some way related. However, the teams had proceeded in an insular manner. Had the connections been identified early on in the project, the teams may have worked together and produced very different findings and recommendations.

Nevertheless, a number of the process evaluations and redesigns were seen as having provided important contributions for eIT's organizational development efforts. In addition, the skills developed during this project were significant for both the students and eIT professionals. Remaining true to the course topic, the most significant benefit for eIT as an organization was the ability to develop collaborative skills. These skills fall into two categories: collaboration with academia and effective collaboration on a decentralized project team. Overall DZI and Temple University viewed this collaboration between industry and academia as a success.

THE OUTCOMES OF THE PARTNERSHIP IN THE EYES OF THE PARTICIPANTS

When asked to rate their agreement with the statement, "Overall, this is one of the best courses I have had at Temple," 58% of the students responded "Strongly agree" (the highest level of agreement); all the remaining students responded "Agree somewhat." The average rating for this question was 3.52 out of 4 (0 = strongly disagree, 4 = strongly agree). Several students pointed out that the course had required much more time and effort from them than traditional courses, which was the reason some students did not rate the course as their "best ever" in the "Strongly agree" category.

A survey was sent to all the eIT project participants and received a 50% return. The results were interesting and suggested a variance in satisfaction between eIT management and eIT staff. While the CIO and DZI's project manager clearly felt that the project was successful as a whole, the eIT staff members differentiated between the content outcomes and the intangible outcomes. Overall the intangible outcomes of exposure to new talent, gaining process design skills, focusing attention on eIT's processes, etc., were ranked high. There was, however, expectation by the process owners that the redesigns would be applied to the processes during the course partnership, which was impossible due to the nature and scope of the process changes, requiring several additional months to be implemented. After all, the process owners had to spend a considerable amount of time and effort in order to participate, and the combination of this with the fact that the process redesigns were not implemented during the course was a source of mild dissatisfaction.

LESSONS LEARNED

Information Sharing Among Teams Does Not Ensure Integration

One of the key lessons learned regarding the use of the Web site is that, even though all the documentation generated by each team was available to all the other teams, process redesign and implementation proposals were developed in relative isolation.

The evidence gathered during the action research study strongly suggests that most teams monitored the work of the other teams through the Web site, by reading team-specific bulletin boards and documents, and even posting comments and suggestions for other teams. However, it seems that the use of that information was restricted to monitoring purposes, so teams would know, for example, if what they were doing was "as good as" what other teams were doing.

This result is in some ways similar to that of an experimental study conducted by Dennis (1996), which found that even though some collaboration technologies may lead users to more information, they may not ensure that the users effectively use the information available. This may be due to information overload (Casey, 1982; Chervany and Dickson, 1974; Kock, 1999a; Meyer et al., 1997; O'Reilly, 1980). That is, even though enough information about the work of each team was available to all teams, their members were not able to effectively process it, probably due to time constraints. This may also explain the fact that even though DZI's CIO had full access to the Web site, he preferred to be briefed about the main outcomes through a face-to-face project review meeting.

The Combined Use of Online and Face-to-Face Interaction Modes is Better Than Having Either Only One or the Other Mode of Communication

The evidence from the action research study suggests that the online interaction is preferable for certain communication and coordination activities than face-to-face interaction, and vice-versa. The combined used of the two modes of communication was seen as a major factor in ensuring the success of the course partnership. The following quote, from a manager directly involved in the partnership, illustrates many of the participants' views regarding this:

It was fantastic how effectively the combination of online discussion and in-person meetings and reviews melded to create a truly collaborative experience…the success of this type of project seems dependent on a blend of both the online and face-to-face interaction. Too much of either would result in a need to extend the schedule, in the case of doing all the work face-to-face, or risking a lack of consensus or true teamwork/team spirit, in the case of a fully electronic experience.

This lesson is aligned with previous research findings (Kock, 1999; Kock and McQueen, 1998). However, the search for optimal combinations of communication modes suggested as relevant by this study is in stark contrast with most of the academic research on collaboration technologies in the 1980s and 1990s, which have focused on experimental comparisons between computer-mediated and face-to-face communication (Kock, 1999).

The Combined Use of Commercial Web Technologies and Interaction Norms Can Remove Computer-Mediated Communication Obstacles

Our study suggests that appropriate use of commercial Web-based technologies can compensate for some of the difficulties inherent in computer-mediated communication. Several empirical studies, particularly those related to media richness theory (Daft and Lengel, 1986; Daft et al., 1987; Kock, 1998; Markus, 1994; Rice, 1992), have shown beyond much doubt that users see certain communication media other than face-to-face interaction as less appropriate for tasks as complex as process improvement. For example, computer-mediated communication is seen as depersonalizing ideas, removing non-verbal cues and preventing immediate feedback, all of which are perceived as having a negative impact on the process and outcomes of teamwork.

On the other hand, prior research findings also suggest that social and organizational norms, such as project guidelines set by management, may compensate for difficulties associated with computer-mediated communication (Markus, 1994). The action research study provides confirmation for this hypothesis. Feedback immediacy, for example, was increased by both the instructor and project manager at DZI directing the participants to check the bulletin boards available from the Web site twice a day and use them as much as possible for interaction regarding the project. As a result, over 300 postings were exchanged within a four-week period, all of which were about tasks related to the course partnership.

In addition to face-to-face meetings and communication behavior guidelines, two commercial Web-based technologies, Internet streaming and image processing, were used to mitigate the depersonalizing effect of computer-mediated communication. Video clips of the CIO, project manager and instructor addressing important issues regarding the partnership were prepared and posted on the Web site along with video clips with team members' introductions. Pictures of the team members, with names added to them, where also posted together with contact information (see Figure 3).

The amount of interaction during the course and the familiarity with which participants from Temple and eIT behaved toward each other during the pizza party is indicative of the "virtual community" sense fostered in part by behavioral norms, electronic interaction and the multimedia components of the Web site.

CONCLUSION

This chapter discussed a course partnership involving Day and Zimmermann, Inc. (DZI), a large engineering and professional services company headquartered in Philadelphia, and Temple University, a public university with its main campus located also in Philadelphia. The course was taught in the First Summer Session of 1999, between the months of May and July. The main goal of the course was to teach students business process redesign concepts and techniques, which were used to redesign five business processes from DZI's information technology organization. DZI's CIO and a senior manager, who played the key role of project manager, championed the course partnership within DZI. A Web site with bulletin boards, multimedia components and static content was used to support the partnership.

Our experience indicates that, given the communication and coordination difficulties associated with such partnerships, the development of a Web site with the features of the one described here is likely to be a key success factor

Figure 3: Images and Video Clips Were Used to Mitigate the Depersonalizing Effect of Computer-Mediated Communication

in similar initiatives. Even though the research literature suggests a number of difficulties associated with conducting projects with the characteristics described, it has been our experience that the combined use of Web-based collaboration technologies with appropriate communication behavior norms and face-to-face interaction is likely to contribute to the success of such projects.

Overall, we believe that the benefits of course partnerships such as the one described here far outweigh their costs. However, we also identified some difficulties that are likely to be faced in similar initiatives. The work of different teams, even when shared among all the participants of the partnership, may not be easy to integrate if synergy is not set as a key goal of the project. Moreover, even though course partnerships may offer several benefits in the eyes of upper management, they may place undesirable pressure on staff who interact directly with students. These problems can be addressed by setting integration of the outcomes of different teams as a key goal of the partnership, as well as building special rewards into the project for the staff involved and those who are able to demonstrate high levels of synergy.

REFERENCES

Brannock, J.C. (1998). Basic guidelines for university-industry research relationships. *SRA Journal*, 30(1/2), 57-63.

Burnham, J.B. (1997). Evaluating industry-university research linkages. *Research Technology Management*, 40(1), 52-56.

Cabral, R. (1998). From university-industry interfaces to the making of a science park: Florianopolis, Southern Brazil, *International Journal of Technology Management*, 15(8), 778-800.

Casey, C.J. (1982). Coping with information overload: The need for empirical research. *Cost and Management*, 56(4), 31-38.

Checkland, P. (1991). From framework through experience to learning: The essential nature of action research. In Nissen, H., Klein, H. K., and Hirschheim, R. (Eds.), *Information Systems Research: Contemporary Approaches and Emergent Traditions*. New York, NY: North-Holland, pp. 397-403.

Chervany, N. and Dickson, G. (1974). An experimental evaluation of information overload in a production environment. *Management Science*, 20(10), 1335-44.

Daft, R.L. and Lengel, R.H. (1986). Organizational information requirements, media richness and structural design. *Management Science*, 32(5), 554-571.

Daft, R.L., Lengel, R.H. and Trevino, L.K. (1987). Message equivocality, media selection and manager performance: Implications for information systems. *MIS Quarterly*, 11(3), 355-366.

Davenport, T.H. (1993). *Process Innovation*. Boston, MA: Harvard Business Press.

Davenport, T.H. and Short, J.E. (1990). The new industrial engineering: Information technology and business process redesign. *Sloan Management Review*, 31(4), 11-27.

Dennis, A.R. (1996). Information exchange and use in group decision making: You can lead a group to information, but you can't make it think. *MIS Quarterly*, 20(4), December, 433-55.

Elden, M. and Chisholm, R.F. (1993). Emerging varieties of action research. *Human Relations*, 46(2), 121-41.

Greenwood, D.J., Whyte, W. F., and Harkavy, I. (1993). Participatory action research as a process and as a goal. *Human Relations*, 46(2), 175-91.

Hammer, M. and Champy, J. (1993). *Reengineering the Corporation*. New York, NY: Harper Business.

Harrington, H.J. (1991). *Business Process Improvement*. New York, NY: McGraw-Hill.

Hollingsworth, P. (1998). Economic reality drives industry-university alliances. *Food Technology*, 52(7), 58-62.

Jones-Evans, D. (1999). Creating a bridge between university and industry in small European countries: The role of the industrial liaison office, *R&D Management*, 29(1), 47-57.

Jonsonn, S. (1991). Action research. In Nissen, H., Klein, H. K., and Hirschheim, R. (Eds.), *Information Systems Research: Contemporary Approaches and Emergent Traditions*. New York, NY: North-Holland, pp. 371-396.

Kock, N. (1998). Can communication medium limitations foster better group outcomes? An action research study, *Information & Management*, 34(5), 295-305.

Kock, N. (1999). *Process Improvement and Organizational Learning: The Role of Collaboration Technologies*. Hershey, PA: Idea Group Publishing.

Kock, N. (1999a). Information overload in organizational processes: A study of managers and professionals' perceptions. In Khosrowpour, M. (Ed.), *Proceedings of the 10th Information Resources Management International Conference*. Hershey, PA: Idea Group Publishing, pp. 313-320.

Kock, N. and McQueen, R.J. (1998). An action research study of effects of asynchronous groupware support on productivity and outcome quality of process redesign groups. *Journal of Organizational Computing and Electronic Commerce*, 8(2), 149-68.

Lau, F. (1997). A review on the use of action research in information systems studies. In Lee, A. S., Liebenau, J., and DeGross, J. I. (Eds.), *Information Systems and Qualitative Research*. London, England: Chapman & Hall, pp. 31-68.

Lee, Y.S. (1998). University-industry collaboration on technology transfer: Views from the Ivory Tower. *Policy Studies Journal*, 26(1), 69-85.

Markus, M.L. (1994). Electronic mail as the medium of managerial choice. *Organization Science*, 5(4), 502-527.

McTaggart, R. (1991). Principles for participatory action research. *Adult Education Quarterly*, 41(3), 168-87.

Mengoni, L. (1998). Cooperation between university and industries in organizing a 'Diploma Universitario' curriculum: The Politecnico di Milano-Assolombarda experience. *European Journal of Engineering Education*, 23(4), 423-30.

Meyer, M.E., Sonoda, K.T. and Gudykunst, W.B. (1997). The effect of time pressure and type of information on decision quality. *The Southern Communication Journal*, 62(4), 280-92.

Meyer-Krahmer, F. (1998). Science-based technologies: University-industry interactions in four fields. *Research Policy*, 27(8), 835-52.

O'Reilly, C. A. (1980). Individuals and information overload in organizations: Is more necessarily better? *Academy of Management Journal*, 23(4), 684-96.

Peters, M. and Robinson, V. (1984). The origins and status of action research. *The Journal of Applied Behavioral Science,* 20(2), 113-24.

Rapoport, R.N. (1970). Three dilemmas in action research. *Human Relations*, 23(6), 499-513.

Rice, R.E. (1992). Task analyzability, use of new media, and effectiveness: A multi-site exploration of media richness. *Organization Science*, 3(4), 475-500.

Saegusa, A. (1997). Japan ties the industry-university knot. *Nature*, 390(6656), 105.

Sommer, R. (1994). Serving two masters. *The Journal of Consumer Affairs*, 28, (1), 170-87.

Wheaton, Q. (1998). Government-university-industry cooperation: Does it work? *Quality*, 37(5), 20-24.

Whyte, W.F. (Ed.) (1991). *Participatory Action Research*. Newbury Park, CA: Sage.

Winter, R. (1989). *Learning from Experience: Principles and Practice in Action-Research*. New York, NY: The Falmer Press.

Wong, P. (1999). University-industry technological collaborations in Singapore: Emerging patterns and industry concerns. *International Journal of Technology Management*, 18(3), 270-85.

Wood-Harper, A.T. (1985). Research methods in information systems: Using action research. In Mumford, E., Hirschheim, R., Fitzgerald, G., and Wood-Harper, A. T. (Eds.), *Research Methods in Information Systems*. North-Holland, Amsterdam, The Netherlands, pp. 169-191.

Section IV

Developing an
IT Curriculum

Chapter X

E-Commerce Curriculum Strategies and Implementation Tactics: An In-Depth Examination of DePaul University's Experience

Linda V. Knight and Susy S. Chan
DePaul University, USA

ABSTRACT

The fast-paced world of e-commerce demands flexible and rapid e-commerce curriculum development. This chapter describes a successful approach to e-commerce curriculum design and development implemented by DePaul University's School of Computer Science, Telecommunications, and Information Systems (CTI). The master's e-commerce curriculum, designed, developed, and implemented in just seven months, drew 350 students in its first year, and approximately 650 students with majors and concentrations in the e-commerce area in its second year. Underlying the curriculum is reliance upon the principles of the IRMA / DAMA 2000, ISCC '99, and IS '97 model curricula. Strong technological expertise and infrastructure, solid industry relationships, and an entrepreneurial culture

Copyright © 2002, Idea Group Publishing.

were critical success factors in developing and implementing the curriculum. The strategies that DePaul CTI employed and the lessons that it learned in the process of implementing its e-commerce curriculum are relevant to other universities seeking to move into the e-commerce arena. Projections are made concerning the future of university programs in e-commerce and the challenges that loom ahead.

INTRODUCTION

In the fast-paced world of e-commerce, where dominant technologies are in a perpetual state of flux and today's hot new business models may be obsolete tomorrow, educators can no longer agonize for years or even months over curriculum decisions. Instead of the measured deliberations of the past, e-commerce demands a rapid and flexible curriculum development that anticipates and keeps pace with ever-changing technologies and business models. This chapter discusses a successful approach to e-commerce curriculum design and development implemented by DePaul University's School of Computer Science, Telecommunications, and Information Systems (CTI). DePaul University, with over 20,000 students, is the foremost granter of master's degrees in the state of Illinois, USA (DePaul, 2001). Known for its entrepreneurial environment, DePaul CTI provides BS, MS, and PhD degrees in a variety of IT-related areas, including information systems, computer science, telecommunications and networking, distributed systems, human-computer interaction, and most recently, e-commerce. Its Master of Science degree in E-Commerce Technology (ECT), with over 400 majors and 650 students total in January 2001, is thought to be the largest e-commerce master's program in the world. While it is difficult to directly transplant any curriculum to another institution, the process that DePaul CTI went through to develop its curriculum, the curriculum design principles and learning objectives that it distilled, the implementation strategies that it employed, and the lessons that it learned all have wide-ranging applicability.

BACKGROUND

The E-Commerce Landscape

E-commerce is broadly defined either as conducting business online, typically by using the World Wide Web, or as applying Internet technologies, especially TCP/IP (the Internet's communication rules) or HTTP (the World Wide Web's communication rules), to business processing. In addition to business-to-consumer (B2C) and business-to-business (B2B) or extranet models, the broad range of e-commerce applications include e-marketplaces where multiple buyers and sellers meet, customer relationship management (CRM), collaborative partnership, information retrieval, electronic publishing, and internal organizational (intranet) applications. These business models and their underlying technologies are emerging and developing as they grow dramatically. Forrester Research (2001) has predicted that total global e-commerce will reach nearly $6.8 trillion in 2004, up from $657 billion in 2000. While the U.S. dot-com euphoria peaked in 2000, the expansion of Internet technology use by traditional brick-and-mortar businesses continues as interest shifts toward bottom-line concerns and system integration.

The rapid growth of e-commerce has triggered demand for e-commerce workers, adding to an existing shortage of technical workers. Even after the dot-com collapse, stock market downturn, and corporate cuts in technology spending during 2000-2001, United States IT salaries remain solid (McGee, 2001), and there is a shortage of over 425,000 IT workers in the U.S. (ITAA, 2001). Further, Web developers head *Computerworld*'s list of the 10 most wanted IT professionals: "In nearly unanimous agreement, recruiters and staffers say that Web developers have quickly surpassed all other job titles in sheer demand. Companies need not just one person but whole teams of Java-experienced developers to design and build the endlessly increasing applications for the Internet" (Bernstein, 2001). In addition to technical skills, hosts of new abilities are required for successful e-commerce development, including graphic arts, user interface design, legal expertise, and writing ability. Web development teams must have both creative and technical leadership, and both of these leadership roles require in-depth understanding of the business and how it functions. Thus, e-commerce demands a broad range of skills and abilities, not just in the IT department, but also throughout the organization. As the worldwide economy becomes more information dependent, demand increases for knowledge workers in a wide range of departments who can facilitate and integrate the use of information throughout the organization.

Universities' Responses

Universities have been criticized for being slow to react to changing conditions in their external environment (Spanier, 2000). According to Harris Miller, president of the Information Technology Association of America, "Relatively few colleges and universities offer a curriculum that directly matches the skills requirements in the IT industry" (Scannell, 1999).

Yet the very nature of e-commerce, an environment that has been characterized as one of "perpetual ambiguity and rapid change" (Nadherny and Stuart, 2000), demands a flexible approach and swift response to curriculum development (Knight and Chan, 2000).

Some universities have indeed responded rapidly to the demand for e-commerce programs. A study of 29 e-commerce degree programs worldwide indicated that business schools are leading the way in e-commerce degrees (Knight, 2001). Approximately two-thirds of the degree programs studied were sponsored by business schools. The rest were either joint efforts of business and technology schools (computer science, information science, or engineering) or the work of technical schools alone. Bachelor's degrees in e-commerce were just beginning to emerge at the time of the study. There were approximately five master's degrees offered for every bachelor's degree in e-commerce found at the institutions studied.

Information Systems Curriculum Models

Since e-commerce is an emerging academic field, there are no existing e-commerce model curricula. However, model curricula can be viewed as generic guidelines for customization (Khosrowpour and Greenawalt, 1997). Thus, existing information systems curriculum models can provide valuable insights into the process and product of e-commerce curriculum design. The three most relevant curriculum models are the Information Resources Management Association (IRMA/DAMA) 2000 Curriculum Model (Cohen, 2000), the IS '97 Curriculum, and the Information Systems-Centric Curriculum '99. Although the three models are designed for the undergraduate rather than the graduate level, they nonetheless isolate numerous key considerations for e-commerce curricula at all levels. All three curricula are listed in the appendix, along with their endorsing organizations and URLs. They have many common elements, yet they differ in terms of emphasis. In the discussion that follows, when one model curriculum is recognized for a particular contribution, this should not be taken to imply that the other curricula do not likewise embrace that concept, but merely as an indication that the first curriculum particularly stresses it.

All three curricula recognize the importance of oral and written communications and team skills. All recognize the increasing role that technology and information play in organizations. All sought and built upon input from industry practitioners, as well as academics, and all have an underlying appreciation of IS as an applied field, inextricably tied to employment. Of the three, the IS '97 curriculum has the longest history, having evolved from a series of curriculum development efforts spread over some 20 years. It represents an attempt to provide a comprehensive program, so that IS graduates will be qualified for employment "across the country," i.e. the

United States. IS '97 defines the IS field as encompassing "acquisition, deployment, and management of information technology resources and services," as well as "development and evolution of technology infrastructures and systems for use in organization processes." The curriculum emphasizes understanding and knowledge in such areas as computer use skills, information systems, computer hardware and software, programming, system development, and database. IS '97 recognizes the need for frequent curriculum updates.

ISCC '99 is heavily focused on industry needs. Not only was this curriculum developed by a collaborative team of industry practitioners and faculty, but industry practitioners set the curriculum's requirements by defining the profile of an IS graduate. Among the ISCC '99 contributions are embedding communication and team skills development within the curriculum, emphasizing programming throughout the curriculum, building in development of student portfolios, and recognizing the key continuing role of industry in maintaining a relevant curriculum and providing opportunities for student learning. According to the ISCC '99 curriculum, "The primary focus is organizing an information system using information as an enterprise asset." This emphasis on the role of information is even stronger in the IRMA 2000 curriculum (Cohen, 2000). This international model curriculum acknowledges information as the most important organizational asset, a strategic and operational resource. The IRMA curriculum is built upon a view of an organization's IS department as not merely a computer system builder and custodian, but a facilitator and integrator of the organization-wide use of information. The IRMA curriculum emphasizes understanding such areas as business systems, organizational behavior, risk management, global issues, and user relationships. When all three model IS curricula are considered, it becomes clear that they each bring unique perspectives to the IS curriculum development process. These perspectives also contribute to the e-commerce curriculum development effort, as described in the next section.

AN EFFECTIVE E-COMMERCE CURRICULUM

The DePaul University Environment

DePaul University is the largest Catholic higher education institution in the United States. Its schools of business and computer technology are located in the heart of Chicago's business district, with eight more DePaul campuses encircling the Chicagoland area. The university values both professional education and technological currency. Its president has made a commitment to overseeing "extensive deployment of technology...to explore new venues of research, learning, and service" (DePaul Minogue, 2001). DePaul's School of Computer Science, Telecommunications, and Information Systems is the largest school of its type in the United States, with approximately 70 full-time faculty and 3,500 students, over 2,000 of them at the graduate level (DePaul CTI, 2001). CTI's technical facilities are state-of-the-art, with specialized student and research labs that include Web Intelligence, Artificial Intelligence, Human Computer Interaction, Collaboration Technologies, Database, Multimedia and Graphics, E-Commerce Technology, Software Development, Local Area Networks, and Telecommunications. CTI's dean has fostered the development of an entrepreneurial environment where faculty members are encouraged to explore new avenues of research and teaching, expanding the school beyond the bounds of traditional Computer Science to include such fields as IS, HCI, and e-commerce. While such a school of technology is not the norm, it is a known model (Knight, 2001).

A significant portion of DePaul CTI's 30% annual growth has been in the areas of IS and e-commerce. In addition to the 405 e-commerce master's majors represented in Figure 1, 240 master's-level IS majors were pursuing concentrations in e-commerce technology at the start of 2001. Complete curriculum requirements, course descriptions, and syllabi for both of these programs, as well as an e-commerce undergraduate degree, are available on the DePaul CTI Web site (DePaul CTI, 2001).

Figure 1: DePaul CTI E-Commerce Technology MS Program Growth

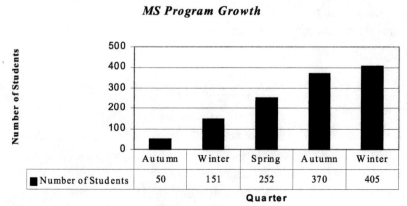

Numbers are approximate

DePaul CTI faculty moved quickly to implement its e-commerce master's program in just seven months. The idea was proposed in February 1999; the design team was formed, background research was conducted, and the curriculum was defined in March; school and university approvals were garnered in April, May, and June. The program was publicized over the summer, and the first 50 students entered in September, just seven months after the project began. The DePaul experience indicates that even a very large university can move quickly and effectively to implement a viable e-commerce program (Knight and Chan, 2000). The remaining sections of this chapter detail DePaul CTI's methodology, and offer insights into the e-commerce curriculum development process and lessons learned.

Curriculum Framework

DePaul CTI's e-commerce curriculum development efforts began with the development of student learning goals and curriculum design principles, coupled with a written definition of the program's target audience. The target audience was defined as those already working within the IT field who wanted to move into a new high-growth area within IT. Such potential students would want to develop the skills and knowledge necessary to work on the technical side of e-commerce, with a career goal of leading e-commerce application develop- ment, either in a large corporation, a consulting firm, or through their own entrepreneurial endeavors. Thus, the DePaul CTI program has a strong appreciation of the importance of technology in e-commerce. Discussions with local employers and anecdotal evidence from students and the DePaul Career Center support the view that technical skills are critical for students hoping to obtain e-commerce employment, but that these skills should be coupled with an understanding and appreciation of business principles, e-commerce busi- ness models, organizational behavior, and the importance and role of the information resource.

Complementing the definition of the degree audience was a set of broad student learning goals. Although this is a more technical e-commerce program, the student learning goals draw from the IRMA/DAMA 2000 curriculum principles (Cohen, 2000) in recognizing the role of MIS as the facilitator and integrator of organization-wide use of information, not merely as the custodian of hardware and software. Notably, the first of these learning goals recognizes the strategic importance of e-commerce to businesses. The full list of goals, developed with both substantial industry input and consideration of the model curricula, appears in Table 1.

Table 1: Student Learning Goals (Knight and Chan, 2000)

Knowledge and understanding of practical applications and evolving e-commerce business models within the context of organizational strategy.
Knowledge and understanding of Web site design principles and engineering processes.
Hands-on experience with e-commerce technologies and tools.
Experience using rapid Web engineering processes to integrate business models, strategies, design methodologies, and technologies.
Experience developing solutions both individually and in a collaborative team setting.

Table 2: Curriculum Design Principles (Knight and Chan, 2000)

Curriculum Design Principle	Reasoning
Expose students to a broad and ever-changing mix of technologies, programming languages, and tools.	Dominant Internet technologies are still emerging and developing.
Build in program flexibility for adapting rapidly to future changes in technologies or e-commerce business models.	The business and technological aspects of e-commerce are changing.
Emphasize practicums and application development for real clients in an authentic e-commerce environment.	E-commerce is, by its very nature, an applied curriculum.
Forge long-term partnerships with e-commerce industry leaders.	Business contacts are invaluable in keeping the program in step with the changing business environment.
Integrate business strategies and development methodologies with technology at the course level; do not segregate business concepts in some courses and technologies in others.	Students should learn e-commerce as they will practice it: as an integrated whole.

Like the learning goals, DePaul CTI's underlying curriculum design principles, listed in Table 2, are also firmly grounded in the principles of IS curriculum models. In particular, the need to emphasize and integrate an appreciation of business strategies throughout the curriculum has direct ties to the IRMA/ DAMA emphasis on information as the most important organizational asset, and the ISCC '99 principle of embedding non-technical skills throughout the curriculum. For the DePaul CTI e-commerce curriculum, this means that while studying the business concepts of supply chain management, a student also studies such underlying business-to-business technologies as messaging and XML, while at the same time building a working extranet within a collaborative team setting. Just as in the business world, there is seamless integration of business strategies, organizational behavior, and operating procedures with technology throughout each course in the curriculum. Knowledge of business principles, managerial skills, systems analysis and design abilities, and technical skills are combined. Business principles include, for example, understanding the role of information as a strategic and operational resource, and knowing how to apply branding and marketing principles to a Web site. Examples of managerial skills include verbal and written communication, team skills, and organizing, planning, and scheduling workflow. Systems analysis and design skills range from designing the user interface, to planning scalable sites and proceeding through the Web development process. Technical skills include, for example, ASP (Microsoft Active Server Pages©) and Java programming, combined with knowledge of commonly used databases and authoring tools. Thus while other universities may teach business principles in some courses and technologies in others, DePaul CTI's e-commerce curriculum integrates business with technology at the course level.

In addition to the integration of business and technology, other DePaul CTI curriculum design principles are also based on previous model IS curricula. DePaul CTI's built-in program flexibility in adapting rapidly to future environmental changes has roots in IS '97's recognition of the need for curriculum updates to keep up with changing technology, and IRMA/DAMA 2000's appreciation of the need to incorporate emerging technologies. E-commerce curricula in particular need to be designed to adapt readily to new technologies, since dominant Internet technologies are still fluctuating as new technologies emerge and mature (Knight and Chan, 2000). DePaul CTI's emphasis upon anticipating industry needs and forging a long-term partnership with e-commerce leaders is also based in prior curricula models, in particular ISCC 99's ongoing collaboration with industry for curriculum updates and development of experiential learning opportunities for students. Finally, the international scope of the DePaul CTI program is consistent with the IRMA/DAMA guidelines.

IMPLEMENTATION STRATEGIES

Course Implementation Strategies

DePaul CTI originally planned to phase-in its new e-commerce degree program over several years. However, due to strong student demand, the full program was implemented in the first year. Initially, the program included new courses such as server-side Web scripting, intranets and business intelligence, and usability for e-commerce. Since most courses were not offered at other universities and course materials were not readily available at the time, the faculty devoted significant effort to curriculum and course development from the start. This process has continued as more advanced courses, such as CRM, Web data mining, e-marketplaces, and mobile commerce, have been added to the curriculum in response to the rapid evolution of e-commerce. Specifically, DePaul CTI employed two noteworthy course implementation strategies:

(1) Implement a Practicum Very Early in the Program

After students complete their prerequisite courses, they integrate their knowledge of business models, value propositions, and Internet technology with Web site engineering projects in a business-to-consumer (B2C) context for the first graduate course of the curriculum. This approach immediately creates a dynamic learning environment with opportunities for students to engage in collaborative processes that carry e-commerce ideas from business development through implementation (Chan and Wolfe, 1999, 2000). Giving students a practicum at the beginning of the program, instead of its conclusion, sets a pace that resembles the dynamic nature of the e-commerce world. Furthermore, students can relate to B2C e-commerce more readily than to the conceptual overview of the Web-based economy that is a common starting point for e-commerce curricula. The CTI approach of starting with a practicum motivates students to acquire development skills and link business strategies with technical solutions early in their studies.

(2) Integrate Business Models and Technology at the Course Level

Since most e-commerce programs are housed in business schools, technology may play a secondary role to business strategies (Malone, 2000; AACSB, 2000). DePaul CTI, as a technically oriented school, faced a somewhat different challenge in identifying appropriate roles for business and technology within its e-commerce curriculum. Instead of stacking together individual business and technology courses, the approach CTI has taken is to consistently integrate business concepts and models, workflows, supporting technologies and tools, and hands-on experiences in each e-commerce course. As a result, students are exposed to a wide range of tools and technologies across the curriculum while learning about the underlying business concepts. Various courses concentrate on both the business and technical sides of B2C, B2B, intranets, CRM, and mobile computing. The only e-commerce course that does not require hands-on development work is e-commerce management, the capstone course emphasizing enterprise e-business strategies. By taking an integrated approach, DePaul CTI allows students to apply a conceptual framework to business process redesign and gain practical experience developing technical solutions for realistic business problems. Celsi and Wolfinbarger (2000) have advocated a related interdisciplinary approach to curriculum for business schools.

Technical Resources and Support Strategies

An effective e-commerce program requires extensive technical re-sources and support. These services must be continuously expanded to meet escalating student and instructional needs. DePaul CTI implemented five technical strategies to meet the demand of the new program:

(1) Build the Initial Technology Support on the Microsoft© Platform

Faced with numerous tools and ever-changing technology, DePaul CTI chose Microsoft© as the common platform for a majority of its courses. Microsoft FrontPage, Access, ASP (Active Server Pages), MSMQ (Microsoft Message Queuing), and IIS© (Internet Information Services) were chosen as the preferred development tool, database, server side scripting tool, message queuing tool, and server platform, respectively. This choice was based in part on the fact that, while local businesses are split between Microsoft© and Unix© platforms, Microsoft's suite was already bundled, interoperable, inexpensive, and easy to learn and maintain. Building on this foundation, more advanced tools, technologies, and languages can be introduced regularly into the pro-gram. Specific technology choices are based heavily in what local industries use. For example, Allaire's HomeSite© was selected as the authoring tool of choice for ASP based upon both an industry survey and the advice of local practitioners. Java was included in the program for similar reasons.

(2) Implement Multiple Logical Servers

Initially, DePaul CTI was concerned about the possibility that student problems in one course might disrupt service for the entire e-commerce program. The school therefore purchased a separate server for each course. However, it became apparent that the servers were more reliable than had been anticipated and that the extra support efforts required for multiple physical servers were not justified. Within the first year, the program switched to just a few larger servers. Approximately 405 e-commerce technology majors and 240 IS majors with ECT concentrations are supported by dual III Xeon 600-megahertz processors, with 1.5 GB of RAM, 50 GB of hard drive space with RAID 1 on system, and RAID 5 on student data drives. The machines have been partitioned into three virtual webs: faculty accounts, student FrontPage accounts, and student FTP accounts. Keeping separate student accounts for the two different approaches to file uploading has significantly reduced server maintenance. Three additional servers—a CRM server, an SQL server, and a MSMQ© server—were made available for advanced courses. The use of multiple logical servers has simplified maintenance and account management for the technical staff.

(3) Build a Dedicated Student Laboratory

A special e-commerce laboratory was built to support this program. This lab is equipped with 333-megahertz Pentium II laptops, each with 96 megabytes of RAM. Laptops were chosen to conserve space for student stations. The computers are arranged in a configuration designed to support either an open student lab or a classroom environment. The room has been equipped with two computer projection systems. An instructor can project the same information on both, or use one to show a Web site while the other shows the code that was used to create it, for example. In addition to the e-commerce laboratory, e-commerce classes are taught in technologically enabled classrooms, complete with a projecting computer and document camera.

(4) Provide a Rich Set of Software Tools

Students in the program learned to use a variety of development and design software, including Microsoft Visual Studio©, Microsoft's Office 2000 with FrontPage©, Allaire's HomeSite©, and Macromedia's Dreamweaver©, Fireworks©, and Flash©, and Adobe's PhotoShop©. For B2B and e-marketplace courses, DePaul CTI experimented with different packages and settled on MSMQ© for messaging, and Brio© and C.5© for business intelligence. Site licenses for these specialized packages are expensive, and their implementation and maintenance on the servers is a substantial investment. It would be possible, however, for universities to implement a respectable program without all of these packages.

(5) Establish a Dedicated Computing Support Staff

An e-commerce program requires expanded technical support, beyond that typically provided by university computing services (Dhamija et al., 1999). At

DePaul CTI, a technical staff of two full-time technicians and several part-time student assistants was created within the school, specifically to support the e-commerce program. This staff is responsible for monitoring and insuring server performance, testing and installation of course software packages, creation and monitoring of student accounts, and setup, monitoring, and maintenance of the student e-commerce laboratory. To support the 700 active student accounts during any given academic term, they have developed account usage policies, online tutorials, and informative Web sites, besides automating the account creation and distribution processes. Instructors and support staff have formed a working group to discuss user and software requirements regularly. Additional support has come from expansion of the existing DePaul CTI tutor corps to include coverage of Web technologies and e-commerce-related skills, including ASP, JavaScript, DHTML, and Java.

Faculty and Staffing Strategies

While the IS faculty at DePaul CTI played a pivotal role in developing and implementing the e-commerce technology program, the new program drew faculty from many disciplines within CTI by using the following six strategies:

(1) Recruit New Faculty Members For Tenure-Track Positions

The school's annual growth rate of 30% insured that sufficient open faculty positions were available. Some new faculty members had prior teaching or research experience in e-commerce, while others relieved existing faculty members, allowing them to move into the e-commerce area.

(2) Collaborate With Faculty in Technical Disciplines

DePaul CTI leveraged the diverse technical background of its faculty in launching the e-commerce technology program. For example, the human computer interaction (HCI) faculty designed and taught usability courses while computer science (CS) faculty members were encouraged to co-design and teach e-commerce courses of a technical orientation.

(3) Emphasize Cross-Disciplinary Retooling

Informal seminars were offered for those CS faculty interested in teaching technical e-commerce courses to prepare them to deal with the organizational aspects of e-commerce and the integrated nature of business, technical, and organizational learning throughout the curriculum. Since the majority of CTI's IS faculty members graduated from business schools, most already had strong business backgrounds as well as technical skills.

(4) Provide Professional Development Opportunities

E-commerce faculty members have unique and challenging professional development needs. The fast-paced, rapidly changing world of e-commerce requires that all faculty members, regardless of background, be constantly involved in learning. Keeping up with a field where changes in both technologies and business models can occur overnight requires a highly motivated, dedicated faculty and ongoing support for their professional development through seminars, training, and opportunities for live business experience and research.

(5) Encourage Sharing of Instructional Materials

When DePaul CTI pioneered the ECT program, relatively few texts were available, and the few available texts became obsolete very quickly. This forced faculty to rely on non-textbook material to an unusual degree. Sharing of instructional materials, many of which are Web-based, became necessary to reduce the burden on faculty.

(6) Actively Involve E-Commerce Professionals in the Classroom

Many consultants and IT professionals were invited to participate as guest speakers and/or adjunct faculty. Their participation has been particularly important for advanced courses that focused on complex e-business solutions.

Partnership Strategies

DePaul CTI leveraged its strong external relationships in the launch of this curriculum through two strategies:

(1) Emphasize Broad Participation of Industry Partners

E-commerce practitioners and consultants have strengthened the program through curriculum development, teaching, guest presentations, student internships, project sponsorship, software support, and roundtable discussions. These professionals and their firms were at the forefront of e-business solution implementation. They were motivated to work closely with universities because of their recruitment needs. Close partnerships with industry have enabled the program to respond to market needs in a timely manner. In addition, industry participation is not exclusive to the e-commerce curriculum. Long before DePaul CTI began development of its e-commerce degrees, it had developed close working relationships with practitioners. Such substantive prior involvement of practitioners provided a rich basis for later curriculum development and was undoubtedly a critical success factor behind the speed of the e-commerce curriculum development process itself.

(2) Leverage Student Involvement

In addition to practitioners, students can also be key to the success of curriculum development efforts. At DePaul CTI, evening master's degree students, working full time in the IT field, provided valuable insights into the specific needs of local businesses for e-commerce professionals. Furthermore, heavy student enrollment in e-commerce pilot classes signaled demand for a full e-commerce curriculum. Student feedback in these courses helped DePaul CTI to gauge the potential demand of the program and make the necessary adjustments to meet student needs.

FUTURE TRENDS

Both the growth of competing e-commerce degree programs during the past few years and the changing e-commerce landscape will put pressure on these programs to undertake continuous curriculum innovation. In the United States, after the 2000-2001 dot-com failures and economic slowdown, continued growth of the Internet economy will be driven by the competitive threat from "efficiency-hungry traditional competitors, not greedy dot-coms" (McCarthy, 2001). As e-commerce efforts increasingly emphasize bottom-line concerns, e-marketplaces and private exchange models are evolving (Haapaniemi, 2001). In some cases, multi-channel strategies and complex system integration have risen to the forefront. Furthermore, the convergence of wireless communications and the Internet will trigger a new wave of mobile business strategies and technology solutions. At the same time, IT professionals are demanding that educators provide more advanced, sophisticated classes (Dash, 2000). These trends will give rise to a need to incorporate more and newer technology into the curriculum, which has clear implications for both faculty development and technology infrastructures (Chan, 2001).

Several trends are emerging from the experience of early adopters of e-commerce curricula (AACSB, 2000; Chan and Knight, 2000; Malone, 2000). Faculty resources and readiness continue to affect the future of e-commerce education. High IS faculty salary costs (Galletta, 2001) may make e-commerce programs difficult or expensive to maintain (AACSB, 2000), forcing universities to consider alternative approaches to staffing. Retooling may be required, not only for business faculty who need to update their technical skills, but also for technically oriented faculty needing business knowledge. Overall, the DePaul CTI e-commerce program's location in a technical school has afforded it many advantages. This experience has shown that "CS faculty bring tremendous technical skills to e-commerce courses but are generally not accustomed to working with business application problems, nor in the more collaborative, team-oriented atmosphere that permeates IS class-rooms" (Chan and Knight, 2000). Business schools considering launching an e-commerce program should consider the possibility of forming strategic partner-ships with their universities' computer science schools. Such a strategic partnership would leverage core competencies on a university-wide basis, providing techno-logical infrastructure and expertise, as well as a source of technically adept faculty. In addition, creative strategies for cross-disciplinary collaboration, team teach-ing, joint course development with industry partners, and systematic faculty development need to be explored.

Effective e-commerce programs demand extensive technology support (Dhamija et al., 1999; Chan and Knight, 2000). This demand increases as a greater emphasis is placed on new technologies with complex integration requirements. DePaul CTI's experience has indicated that consideration must be given to technical support beyond the basic issues of lab operations, access to course servers, and student accounts. The distributed nature of e-commerce solutions, server-side application development, and back-end integration also need to be addressed in a broader context. As an e-commerce curriculum grows, its technical support team may have to operate as an application service provider (ASP), hosting and managing several hundred online quasi-stores and B2B exchanges. A long-term infrastructure plan is a fundamental part of the implementation of any e-commerce curriculum (Chan, 2001).

CONCLUSION

An effective e-commerce program demands rapid implementation of a flexible curriculum. The DePaul CTI experience indicates that critical success factors include strong preexisting industry relationships and an entrepreneurial culture that supports innovation, in addition to strong technological expertise and infrastructure. Business schools may wish to consider forming strategic partnerships with their university's computer science schools to facilitate technology development and supplement faculty resources. Furthermore, any school implementing an e-commerce degree program can benefit from studying the curriculum design prin-ciples inherent in the three existing IS model curricula. Once an e-commerce curriculum is in place, universities still face challenges from accelerating program growth, faculty shortages, and the increasing complexities of e-commerce business models and their supporting technologies. Feed-back mechanisms to insure ongoing curriculum adjustments are critical in sustaining an agile program that is continually in step with the rapidly changing e-commerce landscape.

REFERENCES

AACSB. (2000). Hurr-e up: B-schools striving to get e-business courses and resources up to speed. *AACSB Newsline*, Winter. Retrieved August 1, 2001, from the World Wide Web: http://www.aacsb.edu/publications/newsline/2000/hurre_up_article.pdf.

Bernstein, D. S. (2001). America's 10 most wanted. *Computerworld*, 35(9), 40-41.

Celsi, R. L. and Wolfinbarger, M. (2000). The evolving IT-marketing strategy relationship: Will business schools meet the need? In Chung, H. M. (Ed.), *Proceedings of the 2000 Americas Conference on Information Systems*, 1823-25.

Chan, S. S. (2001). Challenges and opportunities in e-commerce education. *Proceedings of the 2001 Americas Conference on Information Systems*, 1-7.

Chan, S. S. and Knight, L. V. (2000). Implementation strategies for a graduate e-commerce curriculum. In Chung, H. M. (Ed.), *Proceedings of the 2000 Americas Conference on Information Systems*, 1826-1831.

Chan, S. S. and Wolfe, R. J. (1999). Collaborative team learning approach for Web development. *Proceedings of the 1999 Americas Conference on Information Systems*, 181-183.

Chan, S. S. and Wolfe, R. J. (2000). User-centered design and Web site engineering: An innovative teaching approach. *Journal of Informatics Education and Research*, 2(2), 77-87.

Cohen, E. (Ed.). (2000). *Curriculum Model of the Information Resource Management Association and the Data Administration Managers Association*. Retrieved July 30, 2001, from the World Wide Web: http://gise.org/IRMA-DAMA-2000.pdf.

Dash, J. (2000). MBA programs expand e-commerce course offerings. *Computerworld*, August 7, 34(32), 20.

DePaul CTI. (2001). Retrieved March 15, 2001, from the World Wide Web: http://www.cti.depaul.edu.

DePaul Minogue. (2001). Retrieved March 15, 2001, from the World Wide Web: http://www.depaul.edu/visitor/bios/minogue.html.

DePaul University. (2001). Retrieved March 15, 2001, from the World Wide Web: http://www.depaul.edu.

Dhamija, R., Heller, R. and Hoffman, L. J. (1999). Teaching e-commerce to a multidisciplinary class. *Communications of the ACM*, 42(9), 50-55.

Forrester Research, Inc. (2001). Retrieved March 15, 2001, from the World Wide Web: http://www.forrester.com.

Galletta, D. (Ed.). (2001). Association for Information Systems. *2001 MIS Faculty Offer Survey*. Retrieved March 15, 2001, from the World Wide Web: http://www.pitt.edu/~galletta/salsurv.html.

Haapaniemi, P. (2001). The e-market solution. *Chief Executive*, June, 2-4+.

ISCC '99. (1999). *Information Systems-Centric Curriculum '99*. Retrieved March 15, 2001, from the World Wide Web: http://www.iscc.unomaha.edu.

IS '97 Curriculum. (1997). Retrieved March 15, 2001, from the World Wide Web: http://www.is-97.org.

ITAA, Information Technology Association of America. (2001). Executive summary; when can you start; building better information technology skills and careers. Retrieved August 1, 2001, from the World Wide Web: http://www.itaa.org/workforce/studies/01execsumm.htm.

Knight, L. V. (2001). E-commerce curricula: Patterns and projections. In Sagafi, B. (Ed.), *2001 Proceedings; Information Systems and Quantitative Methods, MBAA Conference*, 102-106.

Knight, L. V. and Chan, S. S. (2000). A conceptual model for e-commerce curriculum design and development. In Khosrowpour, M. (Ed.), *Challenges of Information Technology Management in the Twenty-First Century, Proceedings of 2000 IRMA Conference*, 555-557.

Khosrowpour, M. and Greenawalt, D. (1997). The IRM curriculum model: An international curriculum model for a 4-year undergraduate program. *Information Resources Management Journal*, 10(2), 5-20.

Malone, B. (2000). Taking e-commerce to a higher degree. *E-Commerce World Magazine*, December 1. Retrieved March 15, 2001, from the World Wide Web: http://www.ecomworld.com/magazine/issues/article.cfm?ContentID=208.

McCarthy, J. C. (2001). Dot-coms die—but the net economy marches on. *Forrester Research*, June 12.

McGee, M. K. (2001). Salary strongholds. *Informationweek,* 835(April 30), 40-60.

Nadherny, C. and Stuart, S. (2000). The new e-leaders. *Chief Executive*, 59(152), 59.

Scannell, T. (1999). Virtual learning, real students. *Computer Reseller News*, 854 (August 9), 35-36.

Spanier, G. B. (2000). Five challenges facing American higher education. *Executive Speeches*, 14(6), 19-24.

APPENDIX

Information Systems Model Curricula

Curricula	Endorsed by	URL
Information Resources Management Curriculum Model	• Information Resources Management Association (IRMA) • Data Administration Management Association (DAMA)	http://gise.org/ IRMA-DAMA-
Information Systems-Centric Curriculum '99	• Association for Computing Machinery (ACM) • Under consideration by the Institute of Electrical/ Electronic Engineers Computer Society (IEEE/CS) as of September 2000 • Development funded by National Science Foundation	http://www.iscc. unomaha.edu
IS '97 Curriculum	• Association of Information Technology Professionals (AITP, formerly the DPMA or Data Processing Management Association) • Association for Computing Machinery (ACM) • Association for Information Systems (AIS)	http://www.is-97.org

Chapter XI

Information Systems Curriculum Development as an Ecological Process

Arthur Tatnall
Victoria University of Technology, Australia

Bill Davey
RMIT University, Australia

INTRODUCTION

The discipline of Information Systems (IS), in common with the other major branches of computing, is subject to constant and continuing change as new technologies appear and new methodologies and development techniques are devised. IS professionals working in the computer industry need to keep abreast of these changes to remain useful and, of necessity, curriculum in information systems must also undergo frequent revisions and changes.

To those of us involved in research and teaching in information systems, it is clear that curriculum innovation and change in this area is complex, and anything but straightforward (Longenecker & Feinstein, 1990). Of course, all curriculum innovation is complex (Boomer, Lester, Onore, & Cook, 1992; Fullan, 1993; Fullan & Hargreaves, 1992; Kemmis & Stake, 1988) due to the involvement of a large number of human actors, but in information systems curriculum change, this is particularly so, due to the need also to consider the part played by such non-human actors (Latour, 1996) as the technology itself.

We will argue that if you want to understand *how* IS curriculum is built, and how both the human and non-human interactions involved contribute to

Copyright © 2002, Idea Group Publishing.

the final product, then you need to use approaches that allow the complexity to be traced, and not diminished by categorisations (Law, 1999) or assumptions about intrinsic attributes of humans and non-humans. One way that this can be achieved is by using models and metaphors that relate to how people interact with each other, with the environment, and with non-human artefacts. One such approach is provided by the ecological metaphor described in this chapter.

MODELS CURRICULUM DEVELOPMENT

Curriculum change can be modelled in many different ways, and we will here consider just a few of those we consider most relevant. Models of change based upon a process of Research, Development, and Dissemination (RDD) are a common way of attempting an explanation of the process of curriculum development (Nordvall, 1982). In models like this, a rational and orderly transition is posited from research to development to dissemination to adoption.

Although much of the literature relates to the use of these models to explain curriculum change in *schools*, they are also commonly applied to the development of higher education curriculum – the subject of this chapter. Models of this sort suggest that curriculum development follows a logical process of working out the objectives of a particular program, matching these to curriculum elements, developing materials, and then spreading the good word among educators so that the new curriculum will be speedily adopted. We will argue, however, that curriculum change involves a much more complex process than this, and although this approach is one commonly cited in the literature, other models should also be considered. We will now look at four other such models.

An approach that is related to RDD models suggests that many curriculum statements result from the conscious or unconscious copying of 'authoritative' existing statements, rather than from any new thought (Clements, Grimison, & Ellerton, 1989). Although this approach, sometimes known in Australia as the 'Colonial Echo Model,' may have some credence in consideration of curriculum areas such as school mathematics or history, it has been shown to have less relevance in information systems curricula (Tatnall, 1993) at the university level. In most industrialised countries, information systems curriculum was developed primarily in response to local needs (Tatnall, 1993), at least up until the mid-1980s.

Figure 1: Research, Development, Diffusion and Dissemination models

Research ▸▸ Development ▸▸ Production ▸▸ Dissemination ▸▸ Adoption

Since that time, however, well-accepted curriculum documents from groups like IFIP, ACM, DPMA, and IEEE have tended to act as 'authoritative' statements, giving use of this approach more credence today.

A quite different approach makes use of Rapid Application Development (RAD) techniques, adapted from IS systems development. RAD techniques have also been incorporated into a model of IS curriculum development (Davey & Tatnall, 2000). One of the useful ideas to arise from RAD is the concept of a directed process involving 'users' through meetings with very specific agenda, and this can offer advantages in moving forward a curriculum process that has become stalled by directionless debate. The RAD curriculum approach makes use of directed processes involving academics and other stakeholders through a series of meetings with specific agendas.

Information systems curriculum development in a university takes place in the environment of an educational institution containing both academic and industrial elements, and also artefacts such as computers, peripherals, development tools, and methodologies. Any process of curriculum development in such an environment involves a set of complex negotiations between those writing the details of the curriculum. Tatnall (2000) proposes that higher education information systems curriculum can usefully be seen as an actor-network (Callon, 1999; Latour, 1996; Law, 1991) involving the contributions of both human and non-human actors. Approaches such as this see the interactions of individuals and artefacts acting within a system or environment as important, as does the next model we will consider.

This chapter offers the metaphor of an ecological model to help explain how curriculum change occurs within university courses in information systems. Writing about the development of school mathematics curriculum in Australia, Truran (1997) offers what he calls a 'Broad Spectrum Ecological Model' to illustrate curriculum change and to explain the divergence between programs. Truran proposes that systems of education may be seen as ecosystems containing interacting individuals and groups of individuals. The interactions between these parties will sometimes involve cooperation and sometimes competition, and may be interpreted in terms of these interacting forces along with mechanisms for minimising energy expenditure. In this chapter we will examine the application of this metaphor to curriculum change in information systems.

METAPHORS AND MODELS

Before proceeding, however, we need to caution the reader on the limitations and appropriate uses of models and metaphors. A model is an abstraction that is intended as a *representation* of some idea or thing. A model is not itself reality, but just a representation intended to fulfill some explanatory purpose; models thus always have limitations. The dictionary describes a metaphor as a term "applied to something to which it is not literally applicable, in order to suggest a resemblance" (Macquarie Library, 1981:1096).

When a young man says that his girlfriend's eyes glow like diamonds, he is not suggesting that they are made of the hardest material he can think of, but something entirely different. A metaphor is useful, not in giving a literal interpretation, but in providing a viewpoint that allows us to relate to certain aspects of a complex system. In a recent issue of *New Scientist* magazine, James Lovelock, devisor of the Gaia hypothesis, remarked that: "You've got to use metaphor to explain science, it's part of the process of giving people a feel for the subject" (Bond, 2000).

When curriculum developers use the Colonial Echo Model (Clements et al., 1989) of curriculum development, they are using a metaphor that asks us to look for connections between the current curriculum and curricula of the past, or of a 'mother country,' in order to understand the components of the curriculum statement or of the development process. We would contend that most curriculum models and metaphors are too simplistic to allow a useful view of a curriculum, and its development, as a complex system involving human and non-human interactions. An ecological model offers two main advantages:

- A way of allowing for the inclusion of complexity.
- A new language and set of analytical and descriptive tools from the ecological sciences.

AN ECOLOGICAL MODEL OF CURRICULUM CHANGE

In ecology, organisms are seen to operate within a competitive environment which ensures that only the most efficient of them will survive. In order to survive, they behave in ways that optimise the balance between their energy expenditure and the satisfaction they obtain from this effort. These two key principles underlie the discipline of ecology, which is concerned with the relationship of one organism to another and to their common physical environment (Case, 2000; Townsend, Harper, & Begon, 2000). In particular, ecology is concerned with the way that organisms respond to the various forces that operate within the environment.

An ecosystem can be considered to contain producers, consumers, and decomposers. A classical definition is "a natural unit of living and non-living parts that interact to produce a stable system in which the exchange of materials between the living and non-living parts follows a circular path" (Ville,

1962:89). The idea that organisms may be controlled by the weakest link in the ecological chain of requirements, also known as Liebig's Law or the law of the minimum, has been around for some time (Odum, 1963). Under Leibig's Law the rate of growth of an organism is dependent on the nutrient conditions present in the *minimum* quantity in terms of availability and of need.

Habitat, ecological niches, and the exploitation of resources in predator-prey interactions, competition, and multi-species communities (Case, 2000) are all important considerations in ecology. Many different individuals and species typically occupy any given ecosystem, and they can be considered to interact in the following ways:

- When two individuals or species are in **competition** with each other, they are each striving for the same thing, which is typically food, space, or some other physical need. When the thing they are striving for is not in adequate supply for both of them, the result is that both are hampered, or adversely affected, in some manner (Odum, 1963).
- **Proto-cooperation** is the situation in which each population is benefited by the presence of the other, but can survive in its absence.
- **Mutualism** occurs when each population is benefited by the presence of the other, but cannot survive in nature without it (Ville, 1962).
- **Commensalism** occurs when two species habitually live together, and one species is benefited by this arrangement, and the second is unharmed (or not affected) by it.
- The situation in which two species live in close proximity and one species is inhibited by this, but the second is unaffected by the presence of the first, is called **amensalism**.
- When one species adversely affects the second but cannot live without it in the relationship, the result is either **parasitism** or **predation**. These types of interaction are closely related and form a continuum. A small number of predators can have a marked effect on the size of specific prey populations as the energy flow of predators is relatively small.
- The situation in which there is no interaction between two individuals or species is called **neutralism**.

Truran (1997) suggests that these ideas correspond to the process of curriculum development by arguing that an educational system may be seen as an ecosystem, and that the interactions within this ecosystem can be analysed in terms of ecological concepts such as competition models, cooperative behaviour, predator-prey relationships, and niche-development. He argues that curriculum change can be interpreted in terms of interacting forces, mechanisms for minimising energy expenditure, and decisions that individuals make about whether to cooperate or to compete, and goes on to discuss the evolution of high school mathematics curricula in Australia in terms of three ecological principles:

1. **Criteria for behaviour optimisation.** Biological behaviour that requires the least effort to obtain adequate food and shelter is considered optimal. This tendency can be seen in IS curriculum development in examples such as: 'How do we devise a suitable new curriculum in the simplest possible way? Let's just use the material developed by XXXX University.'

2. **Proximate and ultimate behaviour factors.** An ecological example of this is in animal breeding which is seen to be affected by proximate, or short-term, factors such as changes in the hours of daylight, but also by ultimate, or long-term, factors like the need to produce young when food and shelter are most readily available. In relation to IS curriculum, while the need to keep courses relevant and up-to-date with industry developments might be the ultimate aim, proximate factors such as lack of funds to purchase equipment may interfere.

3. **The principle of convergence.** Unrelated families of animals living in the same area are often similar in appearance despite their quite different genetic makeup. This is due to environmental factors. While similarities in curriculum in different countries or locations may seem to suggest that one example has been copied from the other, it is quite possible that they have developed independently due to the similarity in educational environments.

In information systems curriculum development, we should thus look at all the factors, both human and artefact, to see which could be expected to compete, and which cooperate to become part of the surviving outcome. A non-human stakeholder such as a development tool or methodology must cooperate with the environment, compete successfully, or die out. This may mean a new curriculum element becomes incompatible with an old element and so replaces it. Alternatively it may mean that two new design tools can be used together, or that a particular curriculum element is compatible, or perhaps incompatible, with the desires and interests of a particular faculty member.

Ecological metaphors can, however, be used in areas other than just in biology and curriculum change. This type of ecological framework has also been used quite successfully in several other areas including a study of the effects of violence on children (Mohr & Tulman, 2000). Ecology as a framework tells us to expect progress of a task through cooperative or competitive behaviours of the animate and inanimate factors in the environment. A factor that cannot compete or cooperate is inevitably discarded.

CURRICULUM CHANGE AND THE ECOLOGICAL METAPHOR

Many factors influence how an information systems curriculum is developed, and Sandman (1993) outlines these below.

This model provides a good starting point to define the nature of the information systems curriculum ecosystem, and to identify the ecological factors involved in determining how it can be changed.

Ecosystems and Complexity

An ecosystem contains a high degree of complexity due to the large number of creatures and species living in it, and to the variety of interactions possible between each of these. The 'ecosystem' represented by the curriculum in a university information systems department contains (at least) the following 'species': lecturers, researchers, students, professional bodies, university administrators, Course Advisory Board members, and representatives of the computer industry. The 'environment' also contains many inanimate objects relevant to the formation of the curriculum, including: computers, programming languages, textbooks, lecture rooms, analysis and design methodologies, networks, laboratories, programming manuals, and so on.

When using an ecological metaphor, curriculum development can be seen as attempting to introduce change within an ecosystem. The problem of course is the large number of interested parties that must be contended with before this change can be implemented. Curriculum development is more complex than resolving the conflicting needs of students, employers, academics, and the academy. There is ongoing conflict between many things such as educational philosophies, pedagogical preferences, perceived resource constraints, and

Figure 2: Influences on IS Curriculum (adapted from Sandman, 1993

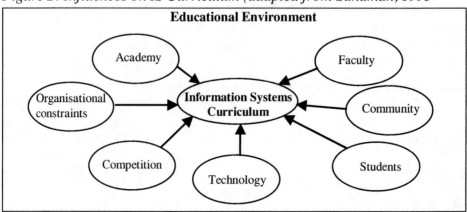

personal issues. Consider the following example of an attempt to introduce a new course in data communications and networking into an undergraduate degree. This would involve many factors and interactions in the ecosystem, including:

- A perception that many job advertisements include the word 'networking' in their text.
- A set of curriculum standards using the word 'networks' to describe a content area.
- A careers adviser who has been asked, by employers, to provide students with more knowledge of networks.
- Several textbooks with the word 'networks' in the title.
- Some students who have expressed an interest in learning about networks.
- An expert teacher who has industrial experience as a network installer.
- An expert teacher who has industrial experience as a network administrator.
- A researcher who has been studying network topologies.
- A researcher who has been studying network software.
- A researcher who has studied the effects of networks on work practices in organisations.
- An Advisory Board member who works for a large bank.
- Another Advisory Board member who works for an IT engineering firm.
- A university facilities manager whose performance is measured by total size of budget and percentage of room usage.
- A curriculum statement aimed at implementing a program that is flexible, and that meets future needs of business.
- A curriculum statement that attempts to give students skills that will be immediately applicable in business.
- A program coordinator who has selected courses that are regarded as essential for an undergraduate education.
- A university ruling that sets the maximum content for a degree.
- Some students who have complained that their program is too technical.
- Some students who have complained that their program is not technical enough.

These are only some of the ecological factors involved in the curriculum development process, and the interactions between these factors will depend upon the power and influence of each factor. We could now ask: Given these factors what will the affect be? Will a networks course be included in the curriculum? If so, will this course involve practical electronics, topological architecture, network administration, or networks in organisations? How long will the course be? Who will be involved with teaching the course? How will the course be assessed? Who will be involved in evaluation of the course?

The answers to these questions, and the progress of the curriculum development, will depend on the complex nature of the ecosystem and the interactions between each of these factors – between the individuals and species in the ecosystem. We will now look at each of the different types of interaction that, apart from neutralism where there is no interaction, can be summarised as: competition, and the various different types of cooperation (proto-cooperation, mutualism, commensalism, and amensalism).

Competition

Competition in nature can occur both within and between species. In many species the males compete with each other for mates, while different species of fish compete for the best feeding areas. In IS curriculum we see many examples of competition, some of which are useful in determining the 'fittest' topics and techniques best suited for survival (Darwin, 1958) in the curriculum, while others involve time-wasting clashes of personality between academics.

One example of competition seen in recent years in many IS Departments is between the programming languages Pascal and Visual Basic. The advocates of Pascal contend that, as a highly structured language, it is still the best vehicle for introducing students to the concepts of programming. Visual Basic advocates, however, argue that while this may be so, Visual Basic is easier to use and moreover is used to a much greater extent in industry. The result of this competition is, most likely, that one language will survive in the curriculum and the other will die out. Similar examples can often be seen in competition between different methodologies and between software products. Most university courses now make use of Microsoft Office rather than Lotus, Work Perfect, and the like, as Microsoft has clearly won the competition and has become dominant in this area.

Sometimes competition can be seen between two teachers with different philosophies or approaches to the teaching of a particular topic area. This can result in the topic being taught twice. Competition between academic departments within a university sometimes results in a topic area being taught in an inappropriate department in order to preserve academic jobs or to retain a balance between the size and importance of these departments. Competition between universities can, unfortunately, result in 'flavour of the month' curriculum development designed with the express purpose of attracting students.

One would hope that predation and parasitism do not have parallels in university IS curriculum development, but most of us know better and could cite an example or two that would fit here also. We will, however, not pursue this line of inquiry here.

Cooperation

There are many examples of unexpected cooperation between organisms in nature: the oxpecker bird that lives with a rhinoceros (commensalism), sharks and suckerfish, barnacles that attach themselves to whales, and dogs and cats living in close proximity with people. It is also possible to think of an organism living in cooperation with its environment: something the native peoples of many countries speak about.

In an educational program such as an information systems degree, some courses rely on earlier courses, i.e., they have prerequisites. This can be seen as a form of cooperation in which each course benefits from the existence of the other (mutualism). Another similar example is in software and programming languages where, for instance, the use of Visual Basic in a computer laboratory requires the presence (and cooperation) of Microsoft Windows. Likewise subject material that relies on the use of a specific textbook could also be seen as an example of cooperation.

Team teaching is an obvious example of cooperation from which all parties benefit. Two teachers working together to develop teaching materials may represent proto-cooperation where both parties benefit, but where either party could get along perfectly well working alone. In many information systems departments, cooperation between teaching and research interests of individual academics results in an improved curriculum.

Ecological Niches

An ecological niche is a place where a particular species that is well suited to this environment is able to thrive, where other species may not. A curriculum example of this is in the teaching of the PICK operating system by a university in Australia. Some years ago PICK was a serious challenger to UNIX for the 'universal operating system' in business, but PICK has now decreased in importance. Despite the fact that no other university in the region now teaches it, and its place being challenged by more recent operating systems, PICK has remained in the curriculum of this university. It has remained largely because an academic involved in its teaching was able to argue convincingly (Tatnall, 2000) that learning PICK allowed students to take up jobs in the small number of prominent local industries using this system – that it filled an important ecological niche.

Other examples include one Australian university that has a close relationship with IBM so that its students work principally with IBM equipment, operating systems, languages, and software. This gives these students an edge in applying for jobs in an IBM environment. Several other universities are working closely with SAP and integrating this product into their courses so that their students gain a better understanding of enterprise resource planning systems and will be able to easily take up jobs in this field.

In each of these cases, the course at the university concerned is aimed at filling a niche – perhaps a small one, perhaps a large one, but a niche nevertheless. The implication is that rather than trying to produce students who know something of the complete field of computing, they will turn out graduates who have specialised in just one or two aspects of it.

IS Curriculum and Other Ecological Processes

Other aspects of ecology can also be related to IS curriculum. Optimisation of behaviour, or minimisation of energy expenditure, is an important principle in ecology as mentioned earlier. It is easy to find examples of this in curriculum development in the use of curriculum templates, and the copying of curriculum from other institutions or countries (Clements et al., 1989). A related example is seen in choosing curriculum elements so that they fit in with existing university resources.

Any factor that tends to slow down potential growth in an ecosystem can be considered as a limiting, or regulatory, factor. The concept of organisms being controlled by the 'weakest link in an ecological chain of requirements' comes from early in the 1800s in Leibig's 'law of the minimum' (Odum, 1963). When any condition approaches or exceeds the limit of tolerance for the organism in question, that is there is too little or too much of this factor, then this will act to limit growth. Many examples of this can be found in IS curriculum change. For example, too little laboratory space or too many students may reduce the quality of learning. Conversely, the availability of a large amount of money for industry-based research may also reduce the amount of academic time spent in preparation for teaching and curriculum development.

Finally, we invite our readers to consider whether they can find curriculum parallels for some of the following ecological situations: island-based examples of evolution such as Australia's specialisation in marsupials, terra-forming by rabbits, herds and pack instincts, and the link between man's introduction of pigs and rats to Mauritius and the extinction of the Dodo.

METAPHORICAL OUTCOMES

To ignore complexity in the curriculum development process will not make it go away, it will just give you less control over this process. We need to decide how to handle complexity and there are different ways that this can be achieved. It would be useful to see how the ecological metaphor might be applied to the process of developing an information systems curriculum, and we will now look at three different ways of thinking about IS curriculum development from the view of this metaphor and from the perspective of three different groups of people. We will call them industrialists, farmers, and biologists.

The curriculum developer who is acting as an *industrialist* works to achieve total control of the educational environment towards the single-minded end of building a curriculum product. In a general context the industrialist has a clear view of what the required outcome is and how it should be manufactured. This may be achieved by activities such as bulldozing the area to be developed, laying concrete, building a factory, and stamping out plastic food. The situation created by the industrialist is a forced one and will only remain in place as long as the industrialist continues to exert his will. In the curriculum context this could be seen to relate to centralised curriculum development and dissemination.

A farmer seeks to exert some control over growing and living things in order to fill some need. He has a clear view of the nature of the crop that is being produced and will seek to exert some control over those parts of the ecology that will lead to a good yield. A farmer will plough the ground and burn the stubble in an attempt to reduce complexity and diversity so that nothing gets in the way of achieving the desired yield. Presuming that he knows what the 'right' crop to grow is, and how to grow it, it will be possible for this crop to remain right over time. A curriculum developer acting as a *farmer* might act, for example, to restrict membership of the curriculum committee to keep out the 'trouble makers,' and to reward cooperative behaviour between groups of academics.

A biologist observes, records patterns, and categorises the behaviour of plants and animals. Standing amidst the complexity of the ecology, the biologist looks for patterns in the interactions between the actors so that useful advice can then be given to the participants. The biologist does not force change, but merely observes it. From her observations she may advise, for instance, that if lions and zebras are allowed to remain in too close proximity, the zebras may not survive. A curriculum developer acting as a *biologist* might advise that if a very strong personality is allowed to dominate curriculum development meetings, then the curriculum is likely to have reduced diversity.

To make use of an ecological metaphor in considering curriculum change, it is first necessary to think of the way you want to view the curriculum development process. We have here offered three ways, but there are no doubt others that could also have been considered. In our example you need to decide if you want to view curriculum as being developed by someone acting like an industrialist, a farmer, or a biologist. When the desired view has been determined, fitting relevant aspects of the metaphor to this view is not a difficult exercise.

CONCLUSION: BENEFITS OF USING AN ECOLOGICAL METAPHOR

Researchers investigating curriculum development, or any other field, must use language in framing their research questions. The language used often reflects a general viewpoint of the field being investigated and will always embody some metaphor for the principle components of the field. The metaphor is not useful in *proving* relationships but can be used to convey meaning once relationships are discovered, and an appropriate metaphor can lead the researcher towards or away from useful possible conclusions. Many of the metaphors for curriculum development are simple ones from areas such as manufacturing or the physical sciences. Any investigation of real development processes in rapidly changing areas such as information systems shows that a common factor is complexity. This leads the search for a suitable metaphor to those disciplines that have accommodated complexity. One such area is that of ecology, and we have shown how ecological principles appear to provide good descriptions of common curriculum development activities. The ease with which the metaphor can be used to describe actions within IS curriculum development shows that it can be useful as a set of language elements that might lead the researcher to framing useful questions that do not trivialise the complexity of the field.

The advantages of the ecological metaphor include the presumption of complexity and interaction. We are not suggesting that the curriculum development process *is* a biological system, but that concepts taken from the field can be seen to be applicable to IS curriculum development. This gives a framework in which researchers can attempt to develop and test models of curriculum development that include the obvious complexity of the real processes.

Many people nowadays tend to use the word 'Web' most often in relation to the Internet, but Darwin (1958) first used it to refer to the set of complex relationships by which plants and animals are bound together, and to their environment in nature. We have argued that this can also be a useful way in which to view information systems curriculum development.

Use of an ecological metaphor provides access to a range of concepts and tools with which to understand the complexities of IS curriculum development and the processes by which this development occurs.

IS curriculum development involves a complex process of negotiation between actors, and one that cannot be simply explained by reference to a set process of referring new ideas to a series of university committees. The choices of individual academics, or groups of academics, to adopt or ignore a new concept or technology, and to compete or cooperate, must also be considered. This inevitably involves a negotiation process between many different actors. We have argued that this negotiation process can be analysed in terms of ecological behaviour, and have utilised an ecological metaphor to assist in visualising the curriculum development process. We have stressed the value of using models and metaphors to describe or illustrate complex activities, and we remind the reader of Lovelock's remark (Bond, 2000) that the value inherent in the use of a metaphor is to give people a 'feel for the subject.' We have argued that use of an ecological metaphor can give people a *better feel* for IS curriculum development.

REFERENCES

Bond, M. (2000). Father Earth. *New Scientist, 167,* 44-47.

Boomer, G., Lester, N., Onore, C., & Cook, J. (1992). *Negotiating the Curriculum: Educating for the 21ˢᵗ Century.* London: The Falmer Press.

Callon, M. (1999). Actor-network theory - The market test. In Law, J. & Hassard, J. (Eds.), *Actor Network Theory and After* (pp. 181-195). Oxford: Blackwell Publishers.

Case, T. J. (2000). *An Illustrated Guide to Theoretical Ecology.* New York: Oxford University Press.

Clements, M. A., Grimison, L. A., & Ellerton, N. F. (1989). *Colonialism and School Mathematics in Australia 1788-1988.* Paper presented at the School Mathematics Conference: The Challenge to Change Geelong, Geelong, Australia.

Darwin, C. (1958). *The Origin of Species.* (Mentor edition.). New York: The New American Library. (Obviously the first edition was published much earlier than this!)

Davey, B., & Tatnall, A. (2000). *Rapid Curriculum Development: A RAD Approach to MIS Curriculum Development.* Paper presented at the Information Systems Education Conference (ISECON 2000), Philadelphia, PA.

Fullan, M. (1993). *Change Forces: Probing the Depths of Educational Reform.* London: The Falmer Press.

Fullan, M., & Hargreaves, A. (1992). Teacher development and educational change. In Fullan, M. & Hargreaves, A. (Eds.), *Teacher Development and Educational Change.* London: The Falmer Press.

Kemmis, S., & Stake, R. (1988). *Evaluating Curriculum.* Geelong: Deakin University Press.

Latour, B. (1996). *Aramis or the Love of Technology.* Cambridge, MA: Harvard University Press.

Law, J. (Ed.). (1991). *A Sociology of Monsters. Essays on Power, Technology and Domination.* London: Routledge.

Law, J. (1999). After ANT: Complexity, naming and topology. In Law, J. & Hassard, J. (Eds.), *Actor Network Theory and After* (pp. 1-14). Oxford: Blackwell Publishers.

Longenecker, H. E. Jr., & Feinstein, D. L. (1990). *Information Systems (IS'90) DRAFT Report: The DPMA Model Curriculum for a Four-Year Undergraduate Degree.* USA: DPMA (CTF-90).

Macquarie Library. (1981). *The Macquarie Dictionary.* Sydney: Macquarie Library.

Mohr, W. K., & Tulman, L. J. (2000). Children exposed to violence: Measurement considerations within an ecological framework. *Advances in Nursing Science, 23*(1), 59-67.

Nordvall, R. C. (1982). *The Process of Change in Higher Education Institutions.* (ERIC/AAHE Research Report 7). Washington DC: American Association for Higher Education.

Odum, E. P. (1963). *Ecology.* USA: Holt, Rinehart and Winston.

Sandman, T. E. (1993). A framework for adapting a MS/MIS curriculum to a changing environment. *Journal of Computer Information Systems, 34*(2), 69-73.

Tatnall, A. (1993). *A Curriculum History of Business Computing in Victorian Tertiary Institutions from 1960-1985.* Geelong, Deakin University.

Tatnall, A. (2000). *Innovation and Change in the Information Systems Curriculum of an Australian University: A Socio-Technical Perspective.* Rockhampton, Central Queensland University.

Townsend, C. R., Harper, J. L., & Begon, M. (2000). *Essentials of Ecology.* Massachusetts: Blackwell Science.

Truran, J. M. (1997). *Reinterpreting Australian Mathematics Curriculum Development Using a Broad-Spectrum Ecological Model.* Paper presented at Old Boundaries and New Frontiers in Histories of Education: Australian and New Zealand History of Education Society Conference, Newcastle, Australia, December 7-10.

Ville, C. A. (1962). *Biology*. Philadelphia, PA: W. B. Saunders Company.

Chapter XII

Information Technology Model Curricula Analysis

Anthony Scime'
State University of New York at Brockport, USA

INTRODUCTION

Most information technology (IT) bachelor degree recipients get jobs after graduation, rather than attend graduate school (Freeman & Aspray, 1999). They enter the workforce because of the tremendous demand for the IT skilled professionals. This means students (and employers) are looking for a practical rather than a theoretical education to fill the computing careers. Such a practical education necessitates a variety of approaches to work in various computing careers. "The traditional career path of programmer to systems analyst to project manager and eventually to IS manager" no longer holds (Urquhart, Perez, Rhoden & Lamp, 1996). With many career paths there is a need for varying academic tracts to start students in their careers.

INFORMATION TECHNOLOGY AS A PROFESSION

Information technology involves the design, development, implementation, support, and management of software and hardware artifacts (Information Technology Association of America, 1997). An artifact may be a chip, a device, a programming language, or a method to store and retrieve data. The jobs done on these artifacts by IT workers can be classified into four categories: conceptualizers, developers, modifiers, and supporters. There is a loose

Copyright © 2002, Idea Group Publishing.

correlation between these categories and IT education (Freeman & Aspray, 1999). It is the job of academia to produce workers in all four of these categories.

Conceptualizers are workers involved with the conception of the basic nature of a computer system artifact. They investigate new ways of processing, storing, transmitting, and representing information. These workers have job titles such as research engineer, systems analyst, computer science researcher, requirements analyst, or system architect.

Developers are people who specify, design, construct, and test an IT artifact. They apply existing technology to new problems. Commonly their job titles are systems designer, programmer, software engineer, computer engineer, chip designer, or tester.

IT workers who modify or extend an information technology artifact work with existing hardware or software. Modifiers maintain systems by making improvements to increase the efficiency of information processing, storage, or communication. They may have job titles such as maintenance programmer, programmer, software engineer, computer engineer, and database administrator.

Finally are those who support or tend the existing systems by delivering, installing, operating, maintaining, or repairing of the information artifacts. Supporters work at the interface between the computer system and the end-user. These are the customer support specialists, help desk specialists, hardware maintenance specialists, network installers, and network administrators.

FORMATION OF INFORMATION TECHNOLOGY AS A DISCIPLINE

Professional fields, such as information technology, are derived from one or more of the traditional disciplines or from other professional fields. The professional fields are the applied application of the knowledge developed in the theoretical disciplines. The professional fields also have an association with the working profession. The working professionals provide guidance concerning the particular knowledge necessary to be successful in the profession. This in turn influences the profession's academic programs.

IT's origins and reference disciplines are management, mathematics, and engineering (Denning, 1998; Freeman & Aspray, 1999; Myers & Beise, 1999; Watson, Taylor, Higgins Kadlec & Meeks, 1999). At any given school, the IT discipline originated from one of these reference disciplines. Regardless of origin the goal of all IT is to process information to be useful.

All academic disciplines are built on a framework of four components. Disciplines have, to varying degrees, components of substantiveness, symbolic systems, syntactic structure, and value/organizational identity. These components form a structure in which the discipline creates a program of study (Dressel & Marcus, 1982). Every program of study must provide students a complete understanding of the organization and functioning of the profession, as well as its knowledge base and techniques (Stark & Lattuca, 1997).

The substantive component of a profession defines the types of problems and the profession's role in society. There are four roles professions fill in society (Stark & Lattuca, 1997). The human-client service role works directly with people by providing a personal service, such as help desk and hands-on support. Information service provides information to others, such as instruction or advice. Professions that build artifacts provide a production service, as in design, development, and programming. Some professions are creative and provide a service by way of artistic expression or entertainment. Of course, some professions possess more than one role. IT provides personal support in the use of computers, advice on the use of technology, and the development of technology.

A discipline's symbolic systems define the language used by the profession to communicate. Languages may include programming languages and models. Professions have two types of symbol systems. Interpersonal symbols allow communication with others outside of the profession. The other communication method is internal to the profession. In IT this is computer code and modeling languages.

The syntactical inquiry methods used by a profession to answer or solve problems are its ways of inquiry. These are the paradigms that define the collection, organization, and analysis of data. Within IT, inquiry is performed by requirements determination and analysis methods.

Concerns of professional identity and ethics are the values of the discipline. This is the relationship of the profession to itself and to society as a whole. The relationship to itself concerns the importance of professional organizations to the individual professional's success in the profession (Stark & Lattuca, 1997). A school's tightness with working professionals manifests itself as formally supervised internships (Stark & Lattuca, 1997; Association of American Colleges, 1992).

IT has all the characteristics of a profession. The founding disciplines of IT are management, mathematics, and engineering. It provides technical service by building, maintaining, and managing hardware systems, software systems, and information systems. Symbolic systems used for discourse in the IT community revolve around mathematical and graphical modeling, programming languages, digital logic, and assembly languages. Inquiry methods into IT require application of requirements analysis, empirical analysis, and experimentation techniques. There are many IT professional organizations. The Association of Computing Machinery (ACM) has a code of ethics, which is accepted as the ethical standard for information technology.

The work categories students can expect to enter may be determined by their school's curriculum. All information technology education consists of various technical, computer-oriented topics ranging from a theoretical understanding of

Table 1: Topics Covered in IT Curricula

IT TOPICS	DEFINITION
Computer Literacy and Use of Software Tools	Practice and theory in the use of personal productivity and personal software development tools to create spreadsheets, simple databases, and other tools for the development of small decision-making systems.
Overview of IT and the Relationship to Business	How information is organized in an information system, how information systems can be used to support business expectations, and the effects of information systems.
Computer Organization and Architecture	Computer organization for the implementation of processors, memory, communications, and interfaces.
Operating Systems and Systems Software	Control mechanisms for the execution of computer programs in an environment of multiple resources.
Programming, Algorithms, Data Structures, & Software Construction	Design and construction of application programs using computer languages to solve mathematical and other models. Includes language syntax, coding styles, and documentation; the organization of data relative to the internals of a program; and the translation of high-level code to machine language for the execution of algorithms.
Networking and Telecommunications	Hardware and software data communication concepts and the design, development, and management of networks.
Systems/Software Requirements Analysis and Design	Requirements collection, analysis, specification; and system design, development, implementation, maintenance, training, extension, and modification.
Database and Information Retrieval	The management of information using a database, the modeling of data relationships, and creation of complex databases using database management systems.
Project Management	The management of complex systems development projects. Issues such as project planning, risk analysis, project administration, and configuration management.
Artificial Intelligence, Robotics, and Decision Support	The design, development, and management of systems that support management, user, or system decisions. Development of models for the construction of machines to simulate human or animal behavior.
Social, Ethical, and Professional Issues	A survey of the legal and moral issues of computer and system usage.
Internship or Design Project	An opportunity to apply classroom knowledge in a professional setting.

computing, through the design and support of practical applications for complex computer systems. These topics collectively represent the body of IT technical knowledge expected of IT graduates with a bachelor's degree as listed in Table 1 (Scime, 2000, 2001). The depth of knowledge in these topics varies with the curricula model.

THE MODEL CURRICULA

The International Federation for Information Processing (IFIP) in coordination with the United Nations Educational, Scientific, and Cultural Organization (USESCO) has developed a framework within which schools can develop an IT curriculum (Mulder & van Weert, 2000). The Informatics Curriculum Framework 2000 (ICF-2000) considers the needs of developing countries for IT-knowledgeable workers. These needs are balanced against the country's educational resources. The result is a tailored IT curriculum based on established models.

The ICF-2000 effort considers information technology education at three levels. The I-user level produces the worker who uses information technology on the job, such as word-processing or simple data entry. The I-applier has the basic IT skills and uses IT to help solve non-IT problems, such as spreadsheet applications. Neither of these is an IT professional. The I-worker is the IT professional that works as a conceptualizer, developer, modifier, or supporter.

The curriculum models to educate IT professionals have been established by the professional societies. Model curricula have also been developed by other organizations, in particular the Software Engineering Institute (SEI) and the Information Systems-Centric Committee (ISCC). These models convey the roles, characteristics, theories, methods, techniques, and practices of the discipline. Each curriculum emphasizes a different perspective of IT. Each fills a different role in providing IT workers.

Information Resources Management Model

IRMA and DAMA view IT as primarily a business function. The preparation, therefore, for IT professionals revolves around the application of IT to business problems. Effective oral and written communication, time management, leadership, and delegation of authority skills, as well as sufficient technical understanding to act as a go-between, expresses this application.

The joint IRMA/DAMA curriculum consists of 10 courses of which the student must complete seven, six required and one elective (Cohen et al., 2000). The management of information and the management, not construction, of computer systems which produce, store, and disseminate information is the curriculum. The curriculum begins after a general business education foundation has been laid (Cohen et al., 2000).

Information Systems Audit and Control Foundation Model

The Information Systems Audit and Control Foundation (ISACF) is the research branch of the Information Systems Audit and Control Association (ISACA). ISACA recognizes that technology has a significant impact on business and that technology must be properly constructed and managed to assure sound business practices. This assurance is maintained through control processes (ISACF, 1998).

This model provides the academic preparation necessary to achieve professional certification as a Certified Information Systems Auditor. The model initially takes a parallel approach to technology and accounting. The proposed curriculum requires 14 courses of which six are information technology, five are accounting and auditing, and two which integrate technology, accounting, and auditing. There are electives in all three fields. The emphasis in technology is in the design and implementation of information systems. This curriculum is designed to fit within a business core of courses (ISACF, 1998).

Organizational & End-User Information Systems Model

The Office Systems Research Association (OSRA) has developed the Organizational & End-user Information Systems (OEIS) model to support the education of the front-line IT professional. The emphasis is on desktop computing. Students completing this curriculum will work with end-users to develop desktop systems that will enhance end-user productivity. A major responsibility of IT support to end-users is training. This is the only curriculum that contains a specific course on how to design and conduct IT training.

The OEIS curriculum consists of seven core courses and nine optional courses (OSRA, 1996). Students are expected to be computer literate when entering the program. None of the optional, elective courses are required to complete the course of study. The elective courses are included to provide specialization in management or design. Because end-users do not program and their supporters work with high-level software development tools, this model has no programming course.

Information Systems '97 Model

The Association for Information Systems (AIS) Model Curriculum for Information Systems, IS' 97, provides the technical aspects in information technology as well as a foundation in business processes. This model is strong in fundamental computing and information technology knowledge. The curriculum is divided into three components. The first level stresses the development of small office and personal systems, the effective use of organizational systems, and the identification of a quality system. The second part specializes in the technology: courses in the hardware and the software of information technology, software programming, and systems analysis and design. Emphasizing teamwork in the design, development, and implementation of information and database system implementation, and project management is the final portion of the course of study (Davis, Gorgone, Couger, Fienstein & Longnecker, 1997). This three level structure allows students to leave the program, having obtained an organized body of knowledge. The curriculum is divided into 10 required technical courses and one technical prereq-uisite. There are no specified technical electives.

Because IS professionals need to be able to communicate within a business organization effectively, courses in communications, quantitative and qualitative analysis, and organizational functions are necessary. Courses outside of IS are also necessary to provide technical background and breadth in communications, mathematics, and business functions (Davis et al., 1997).

Information Systems-Centric Curriculum '99

A collaborative Academe/Industry Task Force is developing a model known as the Information Systems-Centric Curriculum '99 (ISCC '99). This curriculum looks at information as an enterprise asset, which must be managed. This management is accomplished through large-scale, complex information systems that must be built. It is the use of engineering methods to build complex systems that takes precedence in this curriculum.

Industry is a big part of this model's educational process. The last course is a project that comes from an industrial sponsor of the program. The curriculum consists of 12 full courses. Students are encouraged to include in their studies courses that require interpersonal, systemic thinking and problem-solving skills (Lidtke, Stokes, Haines & Mulder, 1999).

Software Engineering Institute Model

The Software Engineering Institute (SEI) considers software development to be an engineering discipline. The curriculum suggested is based upon traditional engineering majors and modeled on the general accreditation requirements of the engineering and computer science accrediting bodies: the Accreditation Board of Engineering and Technology (ABET) and the Computer Science Accreditation Commission of the Computing Science Accreditation Board (CSAB).

The SEI model is a flexible model of software engineering. The knowledge requirement of a software engineer working on operating systems design differs from the knowledge required of an engineer working on an embedded systems project, which differs from the engineer designing a large-scale information system. The SEI model considers these differences and suggests courses be oriented toward a particular type of system within a curriculum. The curriculum consists of nine required courses and unspecified electives in a particular domain (Bagert et al., 1999).

Association of Computing Machinery Models

The ACM and IEEE-CS have jointly published 12 computing curricula models (Tucker et al., 1991). These are known as implementations A-L. However, they can be grouped as Software Engineering, Computer Science, Computer Science in a liberal arts program, and Computer Engineering. The remaining implementations are variations on these themes.

None of the programs has a requirement for business courses. However, in all the implementations there is recognition for the need for humanities, the social sciences, and oral and written communication skills. Mathematics plays a strong, but unspecified, part in the ACM/IEEE-CS curricula.

The Software Engineering program (Implementation G) (ACM-SE) starts early with programming as an engineering discipline. The courses stress software development, large software systems design, programming paradigms, and operating systems. The program culminates with a two-course software design and implementation team project. The curriculum consists of eight required courses and three computer science electives (Tucker et al., 1991).

The Computer Science curriculum (Implementation D) (ACM-CS) has more emphasis on software design. Both hardware and software are covered early. Later courses concentrate on programming languages, operating systems, and computer architecture. There is a six-course requirement for specialization in a sub-field of computer science. The program is capped with a design course. The curriculum consists of 10 required courses, six specialization electives, and the design course (Tucker et al., 1991).

Computer Science in a liberal arts program (Implementation J) (ACM-CS(LA)) presents theory and abstraction early in the program. Later courses discuss specific applications. There is no specialization requirement or design requirement. The curriculum consists of seven required courses and two computer science electives (Tucker et al., 1991).

The Computer Engineering program (Implementation A) (ACM-CE) has a balanced approach to hardware and software. There is a requirement for a specialization in a sub-field of computer engineering, and the program is capped with a two-course design laboratory. The curriculum consists of 12 required courses and six specialization electives (Tucker et al., 1991).

ANALYSIS OF MODELS

Collecting and analyzing data from industry derived all the models reviewed. Industry has expressed a need for emphasis on technical application, interpersonal skills, and business knowledge. These models specifically attempt to reduce problems identified by industry such as a lack of competencies in practical knowledge, inability to work as part of a team, little knowledge concerning the issues of human impact, management of software development, and the economics of software development. All the models recognize the importance of teamwork and effective communication as part of the bachelor's degree program.

Reference Disciplines

From the descriptions of the models, it is clear a model's origin and hence emphasis makes a model closer to one reference discipline than the others. Yet, each model contains some aspects of the other disciplines. The models can be placed in the knowledge space bounded by business, mathematics, and engineering as shown in Figure 1 (Denning, 1998; Freeman & Aspray, 1999; Myers & Beise, 1999; Watson et al., 1999).

Figure 1: Business, Mathematics, Engineering Knowledge Space

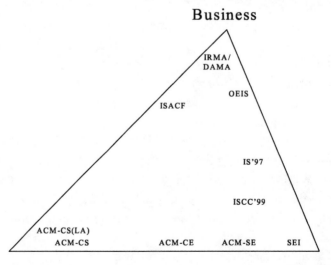

IT Topics and Courses

The courses in each model curricula can be assigned to the IT topics as listed in Table 2. Such an assignment permits a comparison of the models.

Table 2: Number of courses in a topic

TOPIC	IRMA/ DAMA[1]	ISACF [1,7]	OEIS[8]	IS '97	ISCC '99[2]	SEI[3]	ACM-SE[4]	ACM-CS[5]	ACM-CS (LA)[6]	ACM-CE[5]
Computer Literacy and Use of Software Tools		1 R 2 E	1 R 1 E	2 R[9]						
Overview of IT and the Relationship to Business	2 R 2 E	1 E	2 R 1 E	2 R	2 R					
Computer Organization and Architecture				1 R			1 R	3 R	2 R	3 R
Operating Systems and Systems Software							2 R	2 R	1 R	2 R
Programming, Algorithms, Data Structures, and Software Construction	1 R	1 R		1 R	2 R	2 R	2 R	4 R	4 R	4 R
Networking and Telecommunications	1 E	1 R 2 E	1 R	1 R	2 R					
Systems/Software Requirements Analysis and Design	1 R	2 R 3 E	2 R 3 E	2 R	2 R	4 R	2 R	1 R		1 R
Database and Information Retrieval	1 R	1 R 1 E		1 R	1 R					
Project Management		2 E	1 R	1 R		1 R				
Artificial Intelligence, Robotics, and Decision Support	1 R 1 E	1 E	1 E		1 R					
Social, Ethical and Professional Issues			1 E		1 R	1 R				
Internship and Design Project			2 E		1 R	1 R	2 R	1 R		2 R

n R – number of required courses, n E – number of elective courses

1. Selection of one elective required.
2. Four unspecified technical electives required.
3. Unspecified domain electives required.
4. Three unspecified CS electives required.
5. Six unspecified electives in a concentration required.
6. Two unspecified CS electives required.
7. Five accounting and auditing courses required.
8. None of the electives are required to complete the program.
9. One of these is a prerequisite to the program.

Different models place emphasis on different aspects of IT, and therefore will have courses concentrated in different topics. For example, the Systems/Software Requirements Analysis and Design topic has one course from ACM-CS, ACM-CE, and the IRMA/DAMA models; two courses from the IS '97, ISCC '99, and ACM-SE models; four courses from the SEI model; two courses and three electives in the ISACA's ISACF and OSRA's OEIS models; and no courses in the ACM-CS(LA) model.

Curriculum Components and Information Technology Jobs

From the distribution of courses across the IT topics, much can be learned about a model curriculum. The number and type of courses in the model measure the model's component strength as listed in Table 3. For example, in the symbolic component the internal symbol system is composed of programming and modeling languages. Although all the models have a symbolic component, the models that have a more indepth study and applications of IT symbol systems, have stronger symbolic components. The symbolic component can be demonstrated by more than one required course in programming or more than one required course in systems analysis and design. The

Table 3: Component and Role Criteria

Component	Criteria
Substantive Role	
Human Client	At least one required course in computer literacy.
Information	More than one required course in relationship to business.
Production	More than two required courses in systems analysis, databases, project management, or networks, or more than two required courses in programming, or a 2-semester sequence in design.
Symbolic	More than one required course in programming or more than one required course in systems analysis and design.
Syntactic	More than one required course in systems analysis or AI.
Value	At least one required course in ethics.
Organization	At least one required course in design or internship.

Table 4: IT Worker Category Criteria

Category	Criteria
Conceptualizer	More than one required course in relating IT to business or a two-semester design project.
Developer	More than one required course in programming or systems analysis or organization or OS.
Modifier	More than one required course in programming.
Supporter	At least one required course in computer literacy.

professional roles of the substantive component can also be measured by the distribution of courses as listed in Table 3. The human-client service role is one in which the end-user is supported. Courses in computer literacy when taught to IT students should go beyond how-to and into more advanced uses of software tools. This is exactly the focus of such courses in the IS'97, ISACF, and OEIS curricula. (The artistic role is not one typically embodied in IT, and therefore is not considered.)

Likewise, the IT worker categories supported by a curriculum can be determined by the courses as listed in Table 4. Although conceptualizers typically have graduate degrees, at the undergraduate level this thinking can begin by relating IT to the real world, through courses or projects.

The component and worker category criteria can be applied to each model curriculum as listed in Table 5. As can be expected all the models satisfy the production role. How this role is met varies with the symbol system used. The ACM models use the programming courses to introduce the IT symbol systems, whereas the other models use the modeling methods in systems analysis and design courses.

Only required courses are considered in the above strength analysis. Each model also requires some prerequisite work in mathematics. The choice of elective courses can certainly add strength to any of the curricula. The SEI, ACM-CE, and ACM-CS models require that the electives form an area of specialization. The AIS model recommends additional courses in business to complete the curriculum. The ISACF and OEIS have electives beyond the model requirements. Furthermore, courses outside a model's recommendations add variety or focus to each student's program.

Table 5: Model Curricula Strengths

Model	Component Strengths	Substantive Professional/ Service Roles	Worker Categories
IRMA/ DAMA	Syntactic	Information	Conceptualizer
ISACF	Symbolic Syntactic	Human Client Production	Developer Supporter
OEIS	Symbolic Syntactic	Human Client Information Production	Conceptualizer Supporter
IS '97	Symbolic Syntactic	Human Client Information Production	Conceptualizer Developer Supporter
ISCC '99	Symbolic Syntactic Value/Organization	Information Production	Conceptualizer Developer Modifier
SEI	Symbolic Syntactic Value/Organization	Production	Developer Modifier
ACM-SE	Symbolic Syntactic Value/Organization	Production	Conceptualizer Developer Modifier
ACM-CS	Symbolic Organization	Production	Developer Modifier
ACM-CS (LA)	Symbolic	Production	Developer Modifier
ACM-CE	Symbolic Value/Organization	Production	Conceptualizer Developer Modifier

Time

The standard college education is based on an eight-semester sequence of courses. The courses provide for a general education, and specialization is one particular field – the major. The courses in the major are generally interleaved with the general education requirements. Therefore, the number of semesters to complete the major courses may be less than eight as listed in Table 6. The courses in the major are normally arranged is a hierarchical, prerequisite structure (Scime,

Table 6: Number of Courses and Semesters

	Required Courses	Required Prerequisites	Specified Electives	Minimum Semesters to Complete
IRMA/DAMA	7^1	1^2	4	5
ISACF	7^1	7^3	1	5
OEIS	7	—	—	4
IS '97	10	$4^{4,5}$	—	6^5
ISCC '99	12^6	4	4	7
SEI	9	2^5	$—^7$	8^5
ACM-SE	9	1	3	7
ACM-CS	17^1	2	7	7
ACM-CS(LA)	9	1	2	4
ACM-CE	18^1	2	6	8

1. Includes required electives.
2. May be Organizational Behavior or Introduction to Management.
3. These are the non-IT required courses and integrated courses.
4. Specific organizational functional courses are not specified; this table assumes one such course.
5. Includes computing prerequisite.
6. Two ethics courses are half courses and counted as one here.
7. The number of and name of domain electives are not identified.

2000, 2001). This provides for increased depth of knowledge as the student's education proceeds.

IMPLICATIONS FOR THE JOB MARKET

The reason the IT technical societies and others have developed these model curricula is to provide schools guidance in the best way to meet the needs of employers. There is a gap between the needs of industry and the programs provided by four-year schools (Skinner et al., 1999).

Plus, there is a clear lack of skilled professionals to fill the needs of industry. Most of the IT workers are in non-IT companies. One-fourth of all new positions will be in technical support areas. Programming and software engineering is the next most important area for entry-level employees. But, four-year schools generally do not provide adequate background in the technical aspects of networking and telecommunications (Information Technology Association of America, 2001).

Schools that follow one of the models producing workers in the supporter category should have little difficulty in placing their students in jobs. These models are ISACF, OEIS, and IS '97. These support jobs are critical in both IT and non-IT companies. Graduates may be able to find positions throughout the world in these positions.

Developers and modifiers come from the ISACF, IS'97, ISCC '99, SEI, and ACM models. These graduates are well prepared to fill positions in software and database development. Good programmers are needed in the IT companies, which look for entry-level employees with four-year college degrees (Information Technology Association of America, 2001).

All of these models fall short in the area of networking and telecommunications. Five of the models, IRMA/DAMA, ISACF, OEIS, IS '97, and ISCC'99, have networking courses, but these courses take an overview/managerial look at the subject. Technical networking skill knowledge is necessary for the entry-level positions available in industry. Students are forced to find these skills outside the academy in proprietary training institutions and two-year colleges (Information Technology Association of America, 2001; Skinner et al., 1999). This additional training adds time to some students' preparation for the job market.

FUTURE RESEARCH

A long-term study of the match between the needs of industry and the capabilities and success of the various model graduates would be interesting. The distribution of the models among regionally local schools, the types of jobs in that region, and the jobs going unfilled could indicate to a school which model is needed to fill a regional void of IT workers.

The Computing Sciences Accreditation Board (CSAB) already accredits computer science departments. The CSAB is joining the Accreditation Board of Engineering and Technology (ABET), the primary accrediting body for engineering programs in the United States. Under the auspices of CSAB, draft criteria for accrediting information systems programs is in development. IT programs will be accredited by ABET. These accreditation criteria include curriculum requirements (Gorgone & Lidtke, 2000). A look at the IT models in relation to meeting the accreditation standards will also assist schools in selecting an appropriate model.

CONCLUSION

The constantly changing world of computing and the constantly changing world of business leads to the enviable weakness of computing education (Lidtke et al., 1999). All of these models have good points. Schools need to assess their educational philosophy and student needs to choose and modify the most suitable model. By closely following a model, the school's prospective students, students' potential employers, and graduate schools know the type of education received by the graduates. The school administration is assured that the IT department is providing a recognized curriculum, which covers the central IT topics to produce the appropriate graduates.

Although the models differ in emphasis, all businesses that use information (and they all do) will need information technologists from each of the models presented here. There is no one best model. All are necessary, but not all need be provided from the same source.

REFERENCES

Association of American Colleges. (1992). *Program Review and Educational Quality in the Major*.

Bagert, D. J., Hilburn, T. B., Hislop, G., Lutz, M., McCracken, M., & Mangal, S. (1999). *Guidelines for Software Engineering Education Version 1.0.* [Technical Report CMU/SEI-99-TR-032] Software Engineering Institute, Pittsburgh, PA: Carnegie Mellon University.

Cohen, E., Albadvi, A., Alkhatib, G., Aydin, M., Ayre, N., Bialaszewski, D., Childs, R., Christozov, D., Corbin, J., Fedorovich, S., Goldschmidt, P., Gough, T., Harris, R., Janczewski, L., Jentzsch, R., Kamel, S., Kangas, K., Larsen, T. J., Lesjak, D., Nansi, S., Raisinghani, M., Reinhard, N., Schleich, J. F., Soto, L., Swan, J., Targowski, A., Thornburg, G., Warkentin, M., Wigand, R., Xia, G., & Wong, Y. Y. (2000). *IRMA/DAMA Curriculum Model*. Hershey, PA: IRMA. Retrieved November 15, 1999, from the World Wide Web: http://gise.org/IRMA-DAMA-2000.pdf.

Davis, G. B., Gorgone, J. T., Couger, J. D., Fienstein D. L., & Longnecker, H. E. (1997). *IS'97 Model Curriculum and Guidelines for Undergraduate Degree Programs in Information Systems*. Association of Information Technology Professionals.

Denning, P. J. (1998). Computer Science and Software Engineering: Filing for Divorce? *Communications of the ACM,* 40 (8), 128.

Dressel, P. L., & Marcus, D. (1982). *On Teaching and Learning in College*. San Francisco, CA: Jossey-Bass.

Freeman, P., & Aspray, W. (1999). *The Supply of Information Technology Workers in the United States*. Washington DC: Computing Research Association.

Gorgone, J. T., & Lidtke, D. (2000). *Draft Criteria for Accrediting Programs in Information Systems August 2000 (Version 5.2)*. Retrieved June 10, 2001, from the World Wide Web: http://www.csab.org/ docs-pdf/ Draft%20IS%20Criteria%20ver5_2.pdf.

Information Technology Association of America (ITAA). (1997). *Help Wanted: The Workforce Gap at the Dawn of a New Century*. Arlington, VA.

Information Technology Association of America (ITAA). (2001). *When Can You Start? Building Better Information Technology Skills and Careers*. Arlington, VA.

ISACF Task Force for Development of Model Curricula in Information Systems Auditing at the Undergraduate and Graduate Levels, Academic Relations Committee and Research Board. (1998). *Model Curricula for Information Systems Auditing at the Undergraduate and Graduate Levels*.

Lidtke, D., K., Stokes, G. E., Haines, J., & Mulder, M. C. (1999). *ISCC'99 An Information Systems-Centric Curriculum '99: Program Guidelines for Educating the Next Generation of Information Systems Specialists, in Collaboration with Industry*.

Mulder, F., & van Weert, T. (2000). *ICF-2000 Informatics Curriculum Framework 2000 for Higher Education*. Paris: UNESCO.

Myers, M. E., & Beise, C. M. (1999). Recruiting IT faculty. *Communications of AIS*, 2(13). Retrieved November 30, 1999, from the World Wide Web: http://cais.aisnet.org/contents.asp.

Office Systems Research Association (OSRA). (1996). *Organizational & End-User Information System Curriculum Model*. Retrieved December 3, 2000, from the World Wide Web: http://pages.nyu.edu/~bno1/osra/model_curriculum/

Scime, A. (2000). A comparison of information systems model curricula. *Proceedings of the 16th Annual Eastern Small College Computer Conference*. Scranton, PA, pp. 40-53.

Scime, A. (2001). Information systems and computer science model curricula: A comparative look. *Proceedings of the 2001 Information Resources Management Association International Conference*, Toronto, pp. 496-501.

Skinner, R., Abe, N., Adelman, C., Adesegun, N., Barr, A., Barth, J., Brandon, I., Cartwright, P., Cannon, R., Darmody, B., Detweiler, R., Evans, N., Dreyfuss, M., Gallagher, E., Goodwin, K., Gowdy, R., Haines, J., Haverkamp, K. K., Howell, T., Hunt Burris, A., Iskowitz, J., Klima, M., Lafond, J. D., Leffel, L., Lightbourne, J., Livingston, J., Mayo, J. D., Milliron, M., Paddock, R., Pendray, J., Pollard, N., Price, I., Prouty, L., Rowley, T.,

Sargeant, D., Sauer, S., Schmidt, J., Schaefermeyer, M., Smithson-Riehl, J., Sweeney, R., Torrence, M., Valentine, P., Wheeler, S., & Whitaker, R. (1999). *Improving the Responsiveness Between Industry and Higher Education.* Washington, DC: Information Technology Association of America. Retrieved June 20, 2001, from the World Wide Web: http://www.itaa.org/workforce/studies/response.htm.

Stark, J. S., & Lattuca, L. R. (1997). *Shaping the College Curriculum: Academic Plans in Action.* Needham Heights, MA: Allyn and Bacon, pp.113–178.

Tucker, A. B., Barnes, B. H., Aieken, R. M., Barker, K., Bruce, K. B., Cain, J. T., Conry, S. E., Engel, G. L., Epstein, R. G., Lidtke, D. K., Mulder, M. C., Rogers, J. B., Spafford, E. H., & Turner, A. J. (1991). *Computing Curricula 1991: Report of the ACM/IEEE-CS Joint Curriculum Task Force.* Association of Computing Machinery.

Urquhart, C., Perez, A. A., Rhoden, C., & Lamp, J. W. (1996). Professional skills teaching in the IS curriculum: Issues of content, delivery, and assessment. *Proceedings of the Australian Information Systems Curriculum Working Conference*, Melbourne, Australia, pp. 173 -184

Watson, H. J., Taylor, K. P., Higgins G., Kadlec, C., & Meeks, M. (1999). Leaders assess the current state of the IS academic discipline. *Communications of AIS,* 2(2). Retrieved November 30, 1999 from the World Wide Web: http://cais.aisnet.org/contents.asp.

Chapter XIII

Curriculum Model of the Information Resource Management Association and the Data Administration Managers Association

Eli Cohen

Leon Kozminski Academy of Entrepreneurship & Management, Poland
and The Informing Science Institute, USA

EXECUTIVE SUMMARY

Modern organizations recognize the need to maintain and manage information as an organizational asset. They also recognize the need for today's managers to be well versed in information resources management. This document details an international information resources management curriculum for a four-year undergraduate-level program specifically designed to meet needs. The curriculum provides a model for individual universities to tailor to their particular needs. That is, the IRMA/DAMA Curriculum Model is a generic framework for universities to customize in light of their specific situations. This curriculum model prepares students to understand the concepts of information resources management and technologies, methods, and management procedures to collect, analyze, and disseminate information throughout organizations in order to remain competitive in the global business world. These are all aspects of managing information. It outlines core course descriptions, rationales, and objectives, and includes suggested specific course topics and the percentage of emphasis.

Copyright © 2002, Idea Group Publishing.

This curriculum model addresses the needs of two distinct sets of learners:
1.   students currently employed or seeking employment in the IRM field and,
2.   all business students.

The IRM student needs specific in-depth understanding of IRM. All business students, if they are ever to manage effectively, require an understanding of how information management affects their job, the jobs of other managers, and their entire industry.

History of this Curriculum

The IRM Curriculum Model presented here is a revision of two years of extensive research and efforts by an initial joint IRMA and DAMA International Task Force on IRM Curriculum. (DAMA, Data Administration Management Association, International, is an international professional group with more than 3,000 members throughout the world whose emphasis is data and information resource management.) That IRM Curriculum Model began in October 1994. This revision began in 1998 when members of the current task force were invited to bring the curriculum model up to date. This curriculum model was adopted at the IRMA International Conference in Anchorage, Alaska, in 2000.

Relationship to IS '97

Three other IT-related organizations cooperated to produce the IS '97 Curriculum Model (Davis, Gorgone, Couger, Feinstein, & Longenecker, 1997). That model is designed to be generic, and does not specifically meet the needs of IRM programs. In contrast, the IRM curriculum model presented here is designed to match the specific needs for those teaching IRM. Our hope is that schools designing or revising their curriculum will use and benefit from both of these complementary models.

INTRODUCTION

Management information systems (MIS) literature has grown tremendously over the past few decades. Researchers and practitioners alike discuss and assess in this literature new and evolving concepts, applications, problems, and potentials. Both researchers and practitioners recognize the value and importance of MIS in achieving advantage in this very competitive business world. More than ever, effective management of information resources is vital for both national and international firms. More to the point, increasingly firms and governments acknowledge their need for a workforce possessing IRM skills (National Academies, 1999). Also, increasingly, firms recognize that their very survival is dependent on their ability to use information as a strategic and operational resource.

Firms searching for ways to manage their information resources more effectively often discovered that the traditional computer center management lacked the necessary business perspective. Today, top management understands the need for IS personnel who possess a combination of strong managerial skills and strong technical skills.

What does this mean for higher education? It must provide a curriculum that provides students with new skill sets, different from those taught in the past. Educators need to provide future managers with a business orientation by specifically teaching information concepts and theories from the perspective of business problems. The new workplace demands a new curriculum.

A BRIEF HISTORY OF IRM

During the data processing (DP) era of the 1960s, most companies staffed their computer centers solely with people having strong technical backgrounds. Then, the major requirement for managing a computer center was years on the job and technical competence in hardware maintenance and operations. Top management viewed the DP manager position as operational, not managerial. The primary concern for most companies during this DP era was technical feasibility; economic feasibility featured only as a secondary concern.

The DP era tapered off in the 1970s as firms began to apply newly introduced management information systems (MIS) concepts and applications. This new emphasis was on solving managerial problems using technology. The DP departments took on these new tasks, but did not change their staffing. Without acquiring additional skills, former DP managers became the new MIS managers. Unfortunately, the new job required new skills. It also required changes to the organizational structure of the computer center.

The "Fourth Wave" according to Andrews and Carlson (1997) is for the chief executive officer to combine the skills of the technocrat with those of the business executive. Leadership in information systems (IS) and information technology (IT) has changed in fundamental ways over the past decades (Gottschalk, 2000).

Leaders in the field of MIS education recognized the need for MIS professionals to acquire skills, independent of technology, in a wide range of areas including effective verbal and written communication, time management, leadership, and delegation of authority (Metz, Greenhill, & Smith, 1983).

DP -> MIS -> Global Informing Systems

In this decade, the role of the MIS manager has again evolved into a role of managing the information resource. This rapid change places new demands on the education of MIS personnel and managers in general. The orientation of MIS/IRM management now embraces greater user involvement. Consequently, MIS managers, rather than serving as the custodians of the technical computer hardware entities, now function as agents, bringing MIS resources to end-users (Cohen, 1998a).

During this decade, changes to the panorama of technology evoked radical changes in business. This decade witnessed the rise of the Internet, desktop computer, and more affordable telecommunications.

Table 1 shows the components of IRM in modern organizations. It makes clear that the field is comprised of both technical and material components.

ISSUES IN INFORMATION RESOURCES MANAGEMENT

Information resources management (IRM) is the concept that recognizes information as a key resource. A key function of IRM is the study and dissemination of how best to use information to enhance the various organizational functions (Khosrowpour, 1989).

The Value of the Information Resource

One implication of the IRM concept is the need to see the big picture. In the past, people working in information processing departments held the limited view of computer-based information systems comprised solely of its hardware and software. We can blame this limited view in part on limits of their education to the technical aspects. Therefore, today's workers require an increased understanding of the people resources and business applications (Liscouski, 1991).

IRM's integration into all aspects of business during the past few decades created a demand for MIS professionals with abilities not limited to the technical side of information systems. The demand for workers who possess a broad understanding of business systems, organizational behavior, and management began more than a decade ago and has since swelled (Yaffe, 1989; Spruell, 1989). Understanding business systems is important because the integration of IRM into the firm is no longer limited to the formerly technical functions of the firm. At the same time, non-technical workers need

Table 1: Components of Information Resources in Modern Organizations

	Components	Specimen Information Resources Management Issue
Technical	Telecom-munications	Private Branch Exchange (PBX); Local Area Networks; Wide Area Networks; Intranets; Extranets - Internet; Multi-Tier Systems; Virtual Private Networks - Satellite Communication Systems; Facsimile Systems
	Hardware	Legacy Systems; Server-Based Systems; Client/Server Systems; Desktop Publishing Systems
	Software	Operating Systems; Network Operating/Management Systems; Communication Software; Database Management Systems; Object-Relational DB Systems; SQL; Fourth-Generation Languages; Procedural Languages
	Office Automation	Groupware; Electronic Document Management Systems; Electronic Mail; Teleconferencing /Computer/ Video
Managerial	Information System Architectures	Transaction Processing Systems; Management Information Systems; Decision Support System; Executive Information Systems; Expert Systems/Artificial Intelligence - Knowledge Management; Customer Resource Management; Enterprise Resource Planning; Supply Change Integration; Competitive Intelligent Systems
	Information	Marketing; Production; HR Development and Planning Accounting; Legal ; Government; Economy; Financial Strategic; Intelligent; Public; Trade Publications
	Systems Personnel	Programmers; Computer Operators; Data-Entry Personnel; Chief Information Officer (CIO); Systems Managers; Database Administrator/Administration - Systems Security Administrators; Systems Analyst ; Web Master; Web Designer; End-Users
	End-Users	Strategic Planners; Financial Managers; Marketing Managers; Accountants Inventory Managers Production Managers; Personnel Managers; General Managers; Training Personnel; Legal Advisors; Economists
	Management Skills & Programs	Management Styles; Business and IS Strategic Plans; Tactical Plans; Administrative Procedures Training Programs; User Support Programs; Security Procedures and Programs; Disaster Recovery Programs

to hold technical skills. For example, as word processors replaced typewriters, the role of the secretary who typed correspondence evolved into an office information manager. This new role requires greater use of cognitive skills to manipulate and interpret information through word processing, database, desktop, and graphics applications (Regan and O'Connor, 1994).

Organizational and Behavioral Issues

These changes in job demands require changes in education. In particular, today's MIS manager needs to provide more attention to the managerial aspects of IRM. Some MIS managers already find themselves ill-equipped to cope with the behavioral issues that arise in MIS management. More and more workers in organizations have become end-users of computing (e.g., Gasson, 1999; Lehaney, Clarke, Kimberlee, & Spencer-Matthews, 1999).

Integration of IRM into All Functions

Today companies need information systems capable of meeting not only internal company needs, but also the needs of its customers and suppliers. In a given organization, the IS department no longer is strictly a separate function; rather, it has become a function integrated into all departments. The effective MIS manager must have the knowledge to further integrate separate technologies (data processing, telecommunications, etc.) within the organization and to support end-users during this integration (Cohen & Boyd, 1999b).

The integration of technology into the workplace increases security risks, from theft, loss of data integrity including record alterations, and fraud. MIS managers must be knowledgeable in risk management to deal effectively with these risks (e.g., Warden & Trauth, 1999).

Summary

Considering all the issues of information resources management, we have identified the following as critical issues in managing the information resources:

- *Environment*: Today the IS manager must manage a decentralized, end-user-focused environment.
- *Role*: The current IS manager, instead of serving as the technical custodian of computer hardware entities, now functions more like an agent between IS resources and end-users.
- *Expanding Focus*: The IS manager must understand the global issues of the business and its customers, as well as have a comprehensive knowledge of

global information resources management. Information technology has expanded on an international level and, as such, the present focus is now on matters that are more global in nature. The influx of technology into nearly every country has opened a cross-cultural window into other nations that, to this point, was unavailable.

- *Integration*: In a given organization, the IS department is no longer strictly a separate function, rather, it is an integrated function of all departments.
- *Increased Risks*: IS managers must be knowledgeable enough to effectively deal with greatly increased security risks brought about by the integration of technology.
- *Inadequate Preparation*: Business schools continue to graduate students lacking basic knowledge in information resources management.

EDUCATION FOR INFORMATION RESOURCES MANAGEMENT

The Electronic Data Processing (EDP) departments of many businesses came into being during the early 1960s. In response to industry needs, some universities provided courses in data processing (DP), although these early efforts lacked uniformity. EDP courses were taught in departments across campus, from mathematics to accounting. EDP commonly hired bright people with no academic training in EDP and trained them. During this earliest period, programs written for one computer would not "port" to any other.

The main emphasis of these early curricula was on the computing aspects of computer-based systems. These early programs trained many early DP personnel to become programmers and technicians. The academic program of programming and technical knowledge met industry's demands of 30 years ago. Of course, those demands have changed.

Beginning in the late 1960s, as industry moved toward a managerial use of computer-based information systems, colleges began forming new programs in MIS. Professional groups, such as the Data Processing Management Association (DPMA) and the Association for Computing Machinery (ACM), initiated curricula models to assist universities in bringing their diverse programs more in line with the needs of this changing industry.

At that point, businesses viewed their new MIS as a support function. The primary objective of these newly formed MIS programs was to train specialists in programming languages, compilers, and operating systems skills. In addition, something new was beginning to emerge. MIS departments were assisting decision-makers in determining both their information needs and the services that they could obtain from information systems. The age of the systems analyst had arrived.

A quick analysis of the MIS model curricula during the 1970s and early 1980s clearly shows an emphasis on programming and analytical courses. Furthermore, the main emphasis was on hardware and software components of computer-based systems, rather than on information and information users (Mandt, 1982). Even then, proponents of MIS education were recommending the inclusion of courses in oral and written skills, problem solving, and leadership (Metz, Greenhill, & Smith, 1983).

In the 1990s, a new focus evolved in business, the management of information resources—domestically and globally—to benefit organizations. The old needs did not disappear. Firms still needed to manage their information systems. Schools that continued to follow the existing MIS curriculum continued to meet those needs. Nevertheless, they failed to meet the new demands for graduates in information resources management.

This is not to say that MIS programs are flawed. Industry continues to need graduates trained in management information systems. Industries' needs have expanded. Furthermore, the gap between industry needs and education is far from new. In the beginning, industry had to do all the training for their new DP jobs. Even in 1983, tremendous growth in technology and information exceeded the accompanying education, thereby forcing a greater gap between the need to understand and apply information and the ability of the MIS professional to accomplish this task (Porter, 1983). Figure 1 shows the relationship between business processes,

Figure 1: Relationships Between Business Functions, Information Concepts, and Information Technology (Source: Ghazi Alkhatib)

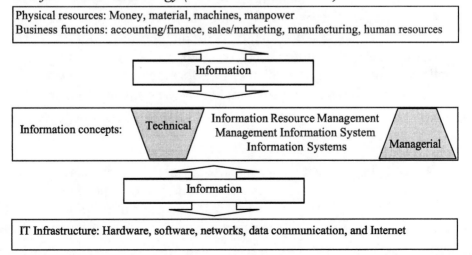

information-related concepts, and information technologies. The figure illustrates that information is the glue between the three components.

Therefore, there are differing needs between these two job functions, and universities' programs should acknowledge these differing needs by establishing separate tracks within their MIS/IRM programs. Commentators acknowledge that MIS education is overdue for a major overhaul to catch up with current trends and demands of the information management age (Ahern, 1992; Laribee, 1992).

Two efforts by professional organizations came out in this period to address the problem. DPMA (renamed later as AITP) and ACM joined forces with the Association for Information Systems (AIS) to create a joint model curriculum, IS '97, which offers the educator a great deal and is a serious contribution. Its strength in attempting to meet the needs of all university programs is also its weakness. It is broad, attempting to contain within its tent program as varied as applied computer science to management of the information system. Consequently, it cannot provide detailed guidance for those developing a program in IRM education. The second effort is this IRMA/DAMA model. It is our hope that educators will benefit from their examination of both IS '97 for its generality and IRMA/DAMA 2000 for its detailed IRM content, as seen below (Srinivasan, Guan, & Wright, 1999).

New Themes in IRM Education

This curriculum model needs to address the needs of the workforce for the twenty-first century. These needs have been examined in-depth and include the following abilities (Cohen & Boyd, 1999; Lin, 1999).

Interpersonal Skills & Communication

Employers ask colleges and universities to provide them with students who possess interpersonal skills and the ability to communicate, both orally and in business writing. This requirement is consistent from study to study and holds equally for information system jobs as it does for others (e.g., U.S. Department of Labor, 1992; U.S. Department of Labor, 1997; Longenecker, Feinstein, Haigood, & Landry, 1999; Lee & Trauth, 1995).

Team Work

These findings above also show that today's jobs require working effectively as members of teams. In the past, IS workers jobs depended purely on technical abilities. Their jobs were conceived as a strictly line position, concerned with the supervision and operations of computer systems, and the preparation of computer-generated outputs that were used predominantly by accountants, controllers, and inventory managers. Today's jobs are for teams of people working in collaboration (Richards, Yellen, Kappelman, & Guynes, 1998).

Understanding Human and Organizational Behavior

In recent years, the role of the IRM/MIS department in business has grown from meeting the needs of only the internal clients to meeting the needs of customers, suppliers, investors, the government, and others. The growth of the Internet and the wide adoption of its technology have spurred this growth. Therefore, our students must now be proficient in many new topics, such as human/computer interaction, understanding customers' needs, legal aspects and people aspects of Web design, and Internet technologies (Stein, Bull, & Burgess, 1995).

Legal, Ethical, and Social Impacts of IRM

Increasingly firms demand that all employees understand the ethical implications of their decisions. In the Longenecker, Feinstein, Haigood, and Landry (1999) study, ethics ranked fourth in importance in a study of what employers want of graduates. (Professionalism was fifth.)

One cannot separate ethics from the rest of one's learning nor from one's culture (Whitman, Townsend, Hendrickson, & Fields, 1998). For this reason, we recommend the study and application of methods for ethical analysis in several courses.

Global Information Issues

The education and training of future MIS managers must ensure that these information experts are capable of developing and implementing effective corporate strategic plans that will succeed in both a national and global business environment. From a government point of view, nations wishing to compete successfully must grow technologically and provide job-oriented education in the field of MIS (Cohen, 1998b; Granger & Lippert, 1998; Deans & Goslar, 1993; Yellen, 1997-98; Cheney & Kasper, 1993).

Research shows an overwhelming consensus reflecting a perspective that: 1) all nations must be prepared to compete in the fast-changing international arena, and 2) education with an international component is a necessary part of the solution (Loch & Khosrowpour, 1993). Some model business schools, including the European Institute of Business Administration (INSEAD) and the International Institute of Management Development (IMD), have already integrated almost twice as much international material into the program as the average business school. Among other innovative ideas, these schools offer programs using overseas training and use international business consultants as classroom instructors (Loch & Khosrowpour, 1993).

There are, however, barriers to developing *globalized* IRM courses. Preparing relevant course materials, training and/or acquiring faculty, obtaining resources, and eliciting organizational support can all prove difficult (Conger, 1993). On some campuses, the development of an internationalized IS curricula has been hampered by the unwillingness of the administration to commit sufficient funding to deploy a program that includes business alliances, international faculty, and overseas experience for students.

Integration of IRM Into All Business Functions

Today companies need information systems capable of meeting not only internal company needs, but also the needs of its customers and suppliers. In a given organization, the IS department no longer is strictly a separate function; rather, it has become a function integrated into all departments. The effective MIS manager must have the knowledge to further integrate separate technologies (data processing, telecommunications, etc.) within the organization and to support end-users during this integration.

Information Security & Quality

One impact of this integration of technology in the workplace is the increase in security risks through theft, loss of data integrity including record alterations, and fraud. MIS managers must be knowledgeable enough to effectively deal with these risks (Applegate, Cash, Mills, & Quinn 1988; Khalil, Strong, Kahn, & Pipino, 1999).

Topical Relevance

We note a strong call for changing what and how we teach to make school more relevant to the job market (Hale, Sharpe, & Hale, 1999; Granger & Lippert, 1998; Burn & Ma, 1997).

PC/Internet/Intranets/E-Commerce

The skill set required of graduates has expanded. Personal computer (PC) and Internet technologies are greatly in demand in today's business. Business's rapid embrace of electronic commerce (e-commerce) means firms look increasingly for our graduates to be prepared in these fields (Athey & Plotnicki, 1998; Slater, 1999).

Introducing IRM Concepts to All Business Students

For some of the resources of productions, business schools have done a good job in educating students: people, money, material, equipment, and management. Shockingly, most fail to adequately educate students to manage information, the emerging key resource of production (Dede, 1989; Jackson, 1992).

One solution to this problem is to teach a separate foundation course in information resources. The IRM foundations course would be similar to principles of marketing or principles of accounting. In it, students would learn about the information resource, its characteristics, utilization, and management.

The current situation at most universities is quite different. Commonly, business students take a computer-skills course as freshmen, followed later by an elective course on IRM/MIS concepts. The message conveyed by this situation is that knowledge of IRM/MIS concepts is less than essential for success on the job. The current situation explains why the employers and governments are calling for revamping business education.

Incorporating IRM Concepts into All Business Courses

A second solution to the problem is to incorporate IRM concepts into all business courses, as suggested by Laribee (1992). In the same fashion that all business courses constantly remind students about the value of financial, human, and other major organizational resources, they need also reinforce across courses the value of the information resources. Commonly, we found poor to no coverage of information resource concepts in any business course. Business programs must guide students to appreciate the value and importance of information resources in all functional areas of a business.

This solution would overcome the danger that students might misperceive an information system as merely a tool to crunch numbers. This proper grounding in the value of information to all business functions will enable students to perceive the essential role of the information resource in today's corporation. In sum, IRM concepts should be incorporated into all business core courses.

IRM as the First Course in MIS Curriculum

The solution above is for all business students. For students majoring in MIS, the first course in the MIS curriculum should introduce students of MIS to the concepts of IRM. Traditionally, the first course in the MIS curriculum focused on applying the tools of information processing systems. We suggest that we should introduce students to information resource concepts before we teach the tools for processing information. Other academic disciplines typically introduce students first to the basic concepts of the field; they teach the tools of the discipline later. For example, in the field of finance, we first teach students the basic concepts of financial resources and only then do we introduce analytical and electronic tools. Students need to understand the problem before they can understand the solution to that problem. Similarly, we must familiarize the MIS student with the value and types of information, characteristics of effective information, users, and sources of information, economy of information, use of it and managerial functions, information and decision-making, information resource management and managers, and the relationship between information resources and other organizational resources.

We must make the concepts and importance of IRM central in the minds of the MIS students before we introduce information processing systems, and particularly before we introduce computer-based information systems. MIS students can become familiar with various information systems and their characteristics and components in a separate course entitled Information Systems Technology after their first course in Information Resource Management. Table 2 provides a summary of the contents of a principle course in IRM.

As with all business students, we should incorporate information management into every course taken in the MIS curriculum. We should teach students skills that will eventually allow them to assist organizations in utilizing their information resources more effectively. For this reason, we need to develop the MIS courses around IRM concepts, rather than information processing concepts.

Table 2: The Primary Components of IRM Concepts

Information management concepts
Data vs. information
Information origins
Types of information
Value of information
Decision making and information
Economy of information
Information processing methods
Information systems structures
Information processing personnel
Information users
Information processing tools
Information processing technology
Information resource managers

The Issue of Texts

The overwhelming majority of these texts are outdated and do not present information management as part of their theme. The textbooks used in MIS courses should cover IRM concepts in depth. Until new textbooks are developed, articles from literature and other supplemental material in IRM should augment current books.

PROPOSED IRM CURRICULUM

What is new in this curriculum model? The traditional MIS curriculum stressed hardware and software, all but ignoring human concerns. The heavy emphasis given to hardware and software over the human components of the systems led in the past to serious problems. (The human component of these systems is not limited to people within the systems, but extends to all end-users and anyone within the organization affected by the information system.) To overcome this limitation of the past, we as teachers must teach the organizational impact of information systems, how these systems fit into the organizational structure of the firm, and how managers can utilize the products of information systems across the company. The first new element to this curriculum is the emphasis on the human elements.

A second element that is new to this curriculum model is the focus on information as a resource. Generally, the model IS curricula of the past has been based on the narrow definition of the computer-based information system as a collection of hardware and software. The critical concepts of IRM outlined here are the foundation of the first two courses in the curriculum: *Information Resources Management Principles* and *Information Systems Technology*. The concepts of IRM are further related to all courses with the MIS program. Other courses offered in this curriculum focus on specific IS topics with supplements of information resources management concepts and applications.

The foundation of any IRM curriculum must begin with an introduction of IRM concepts to all business students, with all subsequent courses building on these concepts to form the curriculum. Critical IRM concepts include:
- The recognition of information as the major organizational asset.
- An understanding of the principles of the characteristics, utilization, and management of information.
- An appreciation of the value and importance of information resources in all functional areas of a business.

- A familiarization with the value and types of information, characteristics of effective information, users and sources of information, economy of information, managerial functions in information-oriented decision-making, and the relationship between information resources and other organizational resources.

Figure 2 illustrates a framework for the previously described IRM course, the information processing technology course, and other courses in the information systems program.

The IRM course is appropriate for university juniors, provided those students have already taken a course in organizational behavior or principles of management course that enables them to understand the organizational and managerial implications of information resource management. The following pages list all of the proposed courses for the IRM curriculum along with course rationale, objectives, and assorted topics. It should be emphasized again that the proposed IRM Curriculum Model should be considered as a generic framework. Adopters should adapt from these guidelines to meet the needs of their local clientele.

The courses in the curriculum can be found on the web at URL http://gise.org/IRMA-DAMA-2000.pdf and for reasons of space cannot be shown here.

Figure 2: An IRM curriculum model for undergraduate education

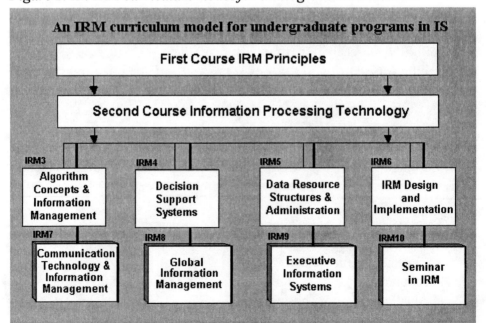

CONCLUSIONS

The realities of the new job market in information management challenges business schools to revise their curricula to provide more education in managing information resources. Graduates of IRM/MIS programs must now have more than merely a command of more than a few programming languages. They must understand the role of information resources in organizations and how to manage these resources effectively. In addition, IRM students need to understand how to provide information to users.

IRM/MIS students should continue receiving adequate training on a wide base of software programs, such as DBMS, expert systems, spreadsheet, and presentation software. Additionally, we need to offer them courses that deal with emerging technologies, applications, and programs using graphics, database strategies, and decision support systems rather than strictly programming languages (Nixon, 1987; Fisher & Hayen, 1990; Jackson, 1992).

The fact that many systems' staff lack a solid educational background in humanistic and organizational concerns points to one of the major reasons for the past and present deficiencies in the management of information resources. This inadequate education can be blamed, in large part, on universities and colleges and their MIS curricula. We are particularly critical of curricula that are exclusively comprised of courses in quantitative analysis and hardware/software-oriented courses, and offer little on the organizational implications of information systems.

IRM/MIS curricula are now due for a major overhaul that will incorporate the concepts of IRM into MIS courses. MIS teachers must keep pace with the demands of corporate America and begin to train information resources managers to understand the value and role of information and to assist organizations in the greater utilization and management of this resource.

IRMA/DAMA 2000 TASK FORCE MEMBERS

Eli Cohen (Chair), Informing Science Institute, U.S. & Leon Komiski Academy of Entrepreneurship and Management, Poland

- Amir Albadvi, Tarbiat Modarres University, Iran
- Ghazi Alkhatib, King Fahd University of Petroleum and Mineral, Dhahran, Kingdom of Saudi Arabia.
- Mehmet Aydin, Boaziçi Üniversitesi, Turkey
- Nicola Ayre, University of Ulster, United Kingdom
- Dennis Bialaszewski, Indiana State University, USA
- Robert Childs, National Defense University, USA

- Dimitar Christozov, American University in Bulgaria, Bulgaria
- James Corbin, Cable and Wireless, Barbados
- Shirley Fedorovich, Embry-Riddle Aeronautical University, USA
- Peter Goldschmidt, University of Western Australia, Australia
- Tom Gough, University of Leeds, UK
- Roger Harris, Universiti Malaysia Sarawak, Malaysia
- Lech Janczewski, University of Auckland, Aotearoa New Zealand
- Ric Jentzsch, University of Canberra, Australia
- Sherif Kamel, American University in Cairo, Egypt
- Kalle Kangas, Turku School of Economics and Business Administration, Finland
- Tor J. Larsen, Norwegian School of Management, Norway
- Duan Lesjak, University of Maribor, Slovenia
- Shi Nansi, Singapore Pools, Singapore
- Mike Raisinghani, University of Dallas, USA
- Nicolau Reinhard, Universidade de São Paulo, Brazil
- John F. Schleich, Assunmption University, Thailand
- Laura Soto, Instituto Tecnológico de Tijuana, Mexico
- Jacky Swan, University of Warwick, UK
- Andrew Targowski, Western Michigan University, USA
- Gail Thornburg, OCLC Online Computer Library Center, Inc., USA
- Merrill Warkentin, Northeastern University, USA
- Rolf Wigand, Syracuse University, USA
- Guoping Xia, Beijing University of Aeronautics and Astronautics, China
- Yuk Yong (Jessie) Wong, Nanyang Technological University – National Institute of Education, Singapore

1995 TASK FORCE MEMBERS

MEMBERS OF IRMA/DAMA CURRICULUM TASK FORCE

- Mehdi Khosrowpour (Chair), Penn State University at Harrisburg, USA
- Abdulla H. Abdul-Gader, King Fahd University, Saudi Arabia
- Anil K. Aggarwal, University of Baltimore, USA
- Thomas Bergin, The American University, USA
- Charles K. Davis, University of Houston-Downtown, USA
- Mary Granger, George Washington University, USA
- Tor Guimaraes, Tennessee Technological University, USA

- Gwynne Larsen, Metropolitan State University, USA
- Tor J. Larsen, Norwegian School of Management, Norway
- Mo Adam Mahmood, University of Texas-El Paso, USA
- Lawrence Oliva, Gale Research Inc., USA
- Prashant Palvia, University of Memphis, USA
- Edward J. Szewczak, Canisius College, USA
- Mohammed Tafti, Hofstra University, USA
- Stu Westin, University of Rhode Island, USA

DAMA TASK FORCE MEMBERS

- Donnette Bruno, DAMA International
- Patricia Cupoli, DAMA International, VP of Education Services
- Gil Laware, DAMA International, VP of Operations
- John Lucas, DAMA International

REFERENCES

Ahern, R. J. (1992). Creating creditable IT curriculum. *Systems Integration Business*, June, 25, 13.

Andrews, P. and Carlson, T. (1997). The CIO is the CEO of the future. *CIO Conference*, Naples, Florida, October 12-15. Retrieved November 4, 1999, on the World Wide Web: http://www.cio.com/conferences/eds/sld01.htm.

Athey, S. and Plotnicki, J. (1998). The evaluation of job opportunities for IT professionals. *Journal of Computer Information Systems*, 38(3), 71-88.

Burn, J. and Ma, L. (1997). Innovation in IT education: Practicing what we preach. *Information Resource Management Journal*, Fall, 16-25.

Cheney, P. H. and Kasper, G. M. (1993). Responding to world competition: Developing the global IS professional. *Journal of Global Information Management*, 1(1), 21-31.

Cohen, E. (1998b). Informing systems for global IRM instruction. In Khosrowpour, M. (Ed.), *Effective Utilization and Management of Emerging Information Technologies*, 949-951. *Information Resource Management Association International Conference Proceedings*, May 17-20. Hershey, PA: Idea Group Publishing.

Cohen, E. (1998a). Something borrowed, something new: The evolution of management information systems. In Callaos, N. (Ed.), *Proceedings of the World Multiconference on Systems, Cybernetics, and Informatics*, 1, 147-152. Orlando, Florida, July 12-16.

Cohen, E. and Boyd. E. (1999a). Reengineering the university–Part II: A plan for overcoming the management information technology labor shortage. *Annual Review of Communications*, 52, 233-236. Chicago, IL: International Engineering Consortium.

Cohen, E. and Boyd. E. (1999b). Reformulating information technology management higher education. In Khosrowpour, M. (Ed.), *Managing Information Technology Resources in Organizations in the Next Millennium*, 97-100. Hershey, PA: Idea Group Publishing.

Conger, S. A. (1993). Issues in teaching globalization in information Systems. In Khosrowpour, M. and Loch, K. D. (Eds.), *Global Information Technology Education: Issues and Trends*, 313-353. Hershey, PA: Idea Group Publishing.

Davis, G. R. (1986). MBA students need more MIS. *Management Information Systems*, 32(19), 19.

Deans, P. C. and Goslar, M. D. (1993). Alterative curriculum approaches for global IT education. In Khosrowpour, M. and Loch, K. D. (Eds.), *Global Information Technology Education: Issues and Trends*, 53-79. Hershey, PA: Idea Group Publishing.

Dede, C. (1989). The evolution of information technology: Implications for curriculum. *Education Leadership*, September, 23-26.

Fisher, E. and Hayen, R. L. (1990). Integrating management information systems and management science: An MBA curriculum response. *The Journal of Computer Information Systems*, 31(1), 35-36.

Gasson, S. (1999). The reality of user-centered design. *Journal of End User Computing*, 11(4), 5-15.

Gottschalk, P. (2000). Information systems executives: The changing role of new IS/IT leaders. *Informing Science: International Journal of an Emerging Transdiscipline*, 3(2), 31-39. Also available on the World Wide Web: http://inform.nu/Articles/Vol3/v3n2p31-39.pdf.

Granger, M. J. and Lippert, S. K. (1998). Preparing future technology users. *Journal of End User Computing*, 10(3), 27-31.

Hale, D., Sharpe, S. and Hale, J. (1999). Business-IS professional differences: Bridging the business rule gap. *Information Resource Management Journal*, April-June, 16-25.

Jackson, D. P. (1992). Curriculum design and the marketplace for MIS professionals. *Interface*, 13(4), 2-7.

Khalil, O., Strong, D., Kahn, B. and Pipino, L. (1999). Teaching information quality in information systems undergraduate education. *Informing Science*, 2(3), 53-59.

Khosrowpour, M. (1988). A comparison between MIS education and business expectations. *The Journal of Computer Information Systems*, Summer, 24-27.

Khosrowpour, M. (1989). Information resources management: A missing course in information systems. *CIS Educator Forum*, 1(4).

Laribee, J. F. (1992). Building a stronger IRM curriculum. *Information Systems Management*, 9, 22-28.

Lee, D. M., Trauth, E. M. and Farwell, D. (1995) Critical skills and knowledge requirements of IS professionals: A joint academic/industry investigation. *MIS Quarterly*, September, 313-340.

Lehaney, B., Clarke, S., Kimberlee, V. and Spencer-Matthews, S. (1999). The human side of information systems development: A case of an intervention at a British visitor attraction. *Journal of End User Computing*, 11(4), 33-39.

Lin, H. (1999). *Workforce Needs in Information Technology*. Retrieved on October 23, 1999, from the World Wide Web: http://www4.nationalacademies.org/cpsma/itwpublic2.nsf.

Liscouski, J. G. (1991). Information management in education. *T.H.E. Journal*, November, 66-69.

Longenecker, H. E. Jr., Feinstein, D. L., Haigood, B. and Landry, J. (1999). IS2000: On updating the IS '97 model curriculum for undergraduate programs of information systems. *Journal of Information Systems Education*, 10(2), 5-7.

Mandt, E. J. (1982). The failure of business education and what to do about it. *Management Review*, August.

Metz, C. H., Greenhill, M. and Smith, R. E. (1983). Preparing DPERS to move up the ladder. *College Studies Journal*, 17, 121-123.

Nixon, J. E. (1987). MIS in a common body of knowledge. *Journal of Education for Business*, 63, 128-130.

Porter, G. L. (1983). The future of MIS education. *Management Accounting*, 64, 14.

Regan, E. and O'Connor, B. (1994). *End-User Information Systems*. New York: Macmillan.

Richards, T., Yellen, R., Kappelman, L. and Guynes, S. (1998). Information systems manager's perception of IS job skills. *Journal of Computer Information Systems*, 38(3), 53-56.

Slater, D. (1999). What is e-commerce? *CIO Enterprise Magazine*, June.

Spruell, J. A. (1989). The MIS domain. *Journal of Education for Business*, 64, 298-302.

Srinivasan, S., Guan, J. and Wright, A. L. (1999). A new CIS curriculum design approach for the 21st century. *Journal of Computer Information Systems*, 39(3), 99-106.

Stein, A., Bull, H. and Burgess, S. (1995). Organizational skill sets for the information professional. *Proceedings of the Inaugural AIS Americas Conference on Information Systems*, August 25-27. Pittsburgh, PA.

U.S. Department of Labor, Employment and Training Administration. (1992). *Learning A Living: A Blueprint For High Performance: A SCANS Report for America 2000*. The Secretary's Commission on Achieving Necessary Skills. Washington, D.C.

U.S. Department of Labor, Employment and Training Administration. (1997). *Involving Employers in Training: Literature Review*. Washington, DC: Research and Evaluation Report Series 97-K.

Warden, F. and Trauth, E. (1999). Balancing the privacy scales: A framework for managing competing values. In Khosrowpour, M. (Ed.), *Managing Information Technology Resources in Organizations in the Next Millennium*, 627-633. Hershey, PA: Idea Group Publishing.

Whitman, M. E., Townsend, A. M., Hendrickson, A. R. and Fields, D. (1998). An examination of cross-national differences in computer-related ethical decision making. *Computers & Society*, 28(4), 22-27.

Yaffe, J. (1989). MIS education: A 20th century disaster. *Journal of Systems Management*, 40(4), 10-13.

Yellen, R. E. (1997-98). A model MBA course in global information systems. *Journal of Computer Information Systems*, 38(2), 41-43.

About the Authors

Eli B. Cohen serves as Professor of Computerized Management Information Systems at the Leon Kozminski Academy of Entrepreneurship and Management (Poland) and Director of the Informing Science Institute (U.S.). He is Co-Founder of the *Informing Science Journal* (http://inform.nu), a refereed academic journal read throughout the world and the *Journal of IT Education* (http://jite.org). He has taught in Australia, Poland, Slovenia, and the U.S. He has formal training and certification in IS, psychology, and education.

Camille Auspitz is Project Manager at Day & Zimmermann, Inc.

Clare Atkins is Academic Leader in the School of Computer and Office Technology at the Nelson Marlborough Institute of Technology and an Honorary Associate of Massey University in Aotearoa/New Zealand. Her research and teaching interests are mainly in the area of relational database design, particularly in requirements acquisition and conceptual modeling and evidence-based information systems. In 1999, Dr. Atkins was recognized by the College of Business at Massey University with two Excellence in Teaching awards, Best Undergraduate Teaching and Overall Distinguished Teacher in the College. The design and delivery of the module described in this chapter was largely responsible for these awards.

Raquel Benbunan-Fich's research interests include educational applications of computer-mediated communication systems, asynchronous learning networks, and evaluation of commercial Web-based systems. She has published articles on related topics in *Group Decision and Negotiation, Journal of Computer Information Systems,* and *Journal of Computer-Mediated Communication*, among others, and has forthcoming articles in *Communications of the ACM* and *Information & Management.* Dr. Benbunan-Fich is an Assistant Professor of Information Systems at the W. Paul Stillman School of Business, Seton

Copyright © 2002, Idea Group Publishing.

Hall University. She received her PhD in Management Information Systems from Rutgers University in 1997.

John Bentley is a Senior Lecturer in Information Systems at Victoria University of Technology in Melbourne. He is the Course Director of the Bachelor of Business in Computer Systems Support conducted in Melbourne and Hong Kong. He has coordinated the activities of the Faculty of Business on the Melton Campus in the outer western suburbs of Melbourne. Professor Bentley has been a member of the Faculty Board of Studies and a foundation member of the Academic Board of the university. He is currently undertaking a PhD concerning the curriculum of information systems undergraduate courses and graduate attributes. He has a strong teaching background combined with relevant industry experience and consultancy work. His research interests include small business and electronic commerce, Web usability, user issues in systems development, work-integrated learning, and problem-based learning.

Susy S. Chan is an Associate Professor and the Director of the Center for E-Commerce Research in the School of Computer Science, Telecommunications, and Information Systems, at DePaul University. She was the Co-Creator of DePaul University's pioneering master's and baccalaureate programs in e-commerce technology. As a former CIO and Vice President for Planning at DePaul, she developed its six-campus IT infrastructure. Her research is focused on e-commerce development and strategies, mobile commerce, enterprise transformation, and ERP procurement. She received a PhD in Instructional Technology from Syracuse University.

Chris Cope has been teaching about information systems and researching into IS teaching and learning for 13 years. He is currently an academic in the Division of IT at La Trobe University, Bendigo, Australia.

Bill Davey is a Senior Lecturer in the School of Business Information Technology at RMIT University, Melbourne, Australia. He holds bachelor's degrees in Science and Education and a master's of business degree. His research interests include methodologies for systems analysis and systems development, Visual Basic programming, information systems curriculum, and information technology in educational management. He & Arthur Tatnall have worked together and cooperated on many occasions. They have worked on joint research projects and co-authored a large number of papers, book chapters, and textbooks relating to management information systems, programming, and IS curriculum.

Mariam Fergusson works at PricewaterhouseCoopers in Global Risk Management Solutions focusing on e-business and information technology security in the public sector. She was a Senior Lecturer in the School of Computer Science, UNSW, at the Australian Defence Force Academy in Canberra from 1995-2000. The focus of her teaching was to postgraduate students in information systems research methods and telecommunications. During this time, she undertook significant research into the teaching of postgraduate research students. Prior to that, she was the coordinator of an MBA program in the United Arab Emirates. Ms. Ferguson's research interests are in electronic commerce, particularly in public sector service delivery. Over the last three years, she has worked with Australian federal and state government agencies on a range of electronic business projects including Internet payment systems, and has published a book chapter and several articles on this topic. She has worked in the information technology industry for the last 20 years, dividing her time between consultancy work, and an educator and researcher at universities.

Beverley G. Hope is employed as Teacher and Researcher in the School of Information Management at Victoria University of Wellington, New Zealand, where she became involved in the planning for a new Honours and Research Master's program in 1997. The programme was offered for the first time in 1998. During the completion of this chapter, Dr. Hope was supported by a Visiting Fellowship at the City University of Hong Kong. She completed a Bachelor's of Science and MBA at the University of Kansas and a PhD at the University of Hawaii at Manoa. Her PhD studies were supported by a scholarship from the East-West Center. Her research and teaching focus on quality, information management, current IS issues, and IS research. In her work she takes a holistic or system view of organisations and the issues that face them. She has presented and published at many regional and international conferences, and acts as referee for several international conferences and journals.

Pat Horan is a Senior Lecturer in the Department of Information Technology at La Trobe University, Bendigo. After programming in the days when mainframe Cobol was king, she moved into tertiary IT teaching, evolving through technical and advanced education colleges to university. Her current research interest is how students learn about IS and her ongoing PhD study is in this field. In addition to students her interests include animals (especially dogs, cats, goats, and horses), opera, cryptic crosswords, films, red wine, and bemused observation of the antics of politicians and university management.

Brad King is Chief Information Officer at Day & Zimmermann, Inc.

Linda V. Knight is Associate Dean of DePaul University's School of Computer Science, Telecommunications, and Information Systems. She is also Director of CTI's Center for the Strategic Use of Emerging Technologies. Co-developer of one of the world's largest e-commerce degree programs, she lectures on the topic of e-commerce curricula and teaches and conducts research in the area of e-commerce business strategy, development, and implementation. She is Associate Editor of the Journal of IT Education, and services on the Editorial Review Board of the Information Resources Management Journal. An entrepreneur and IT consultant, she has held industry positions in IT management and quality assurance management. She holds a PhD in Computer Science from DePaul University.

Ned Kock is the Director of the E-Collaboration Research Center, Fox School of Business Management, Temple University, Philadelphia, USA. He is also a CIGNA Corporation Research Fellow and an Irwin L. Gross E-Business Institute Research Fellow.

Glenn Lowry is Professor and foundation Executive Director of MBA Programs at United Arab Emirates in Abu Dhabi. He earned the PhD in Information Systems from Rutgers University, USA. He served in senior academic and administrative posts at Virginia Tech, the University of Iowa, University of Technology, Sydney, the University of Tasmania, and as Visiting Research Fellow at Victoria University. He has also worked in industry as an account manager at Technical Aid Corporation in Washington, DC, and has consulted to NCR, Fujitsu, and Sumbershire Management (Singapore). He has lectured throughout the U.S., Australasia, Europe and the UK, Africa, and the Middle East. With research interests in technology-led innovation and change management, cross-cultural technology acceptance, software engineering, e-business, research methods, and information systems education, Dr. Lowry has authored 60+ papers and six books.

Terry D. Lundgren is a Professor of Information Systems at Eastern Illinois University where he has been on the faculty since 1989. His research and writing have resulted in more than 100 publications, including co-authoring *Records Management* and *Advanced Microcomputer Applications* textbooks. His current research interests include enhancing microcomputer and programming courses by Web technology. He has been an advocate of computer use since the early 1970s and is on technology's leading edge in course development and delivery.

Geoffrey C. Mitchell is employed as a Senior Lecturer in the University Teaching Development Centre at Victoria University of Wellington, New Zealand. After six years teaching in the information systems area, Dr. Mitchell moved to the UTDC to support the introduction of new courseware systems across the university. He has been the recipient of innovation in teaching awards and research grants for his work in the development of online teaching systems. His PhD research in Australia and New Zealand has focused on the politics of new systems introduction in public sector organisations.

Karen S. Nantz is a Professor of Information Systems at Eastern Illinois University where she has been on the faculty since 1990. Dr. Nantz is the co-author of *The 3 R's of Email: Risks, Rights, and Responsibilities*, along with numerous articles in academic journals. She currently serves as an Associate Editor of the *Journal of Information Technology Education* and is on the editorial boards of the *Annals of Cases on Information Technology Applications* and *The Journal of End User Computing*. Her teaching interests include systems analysis, database, and management of information systems.

Geoff Sandy has worked in tertiary education for 25 years and the private sector for five years, first at BHP Petroleum and then the ANZ Banking Group. His main research interests are learning strategies, censorship of the Internet, and network usage policy. He has published widely in these areas, teaches at the undergraduate and postgraduate levels, and supervises research degree students. He is a member of the Academic Board of Victoria University, Australia, and the Committee for Postgraduate Studies and the Resources Planning and Advisory Committee. Dr. Sandy has provided consultancy services over many years to a diverse group of organizations from the private and public sectors.

Anthony Scime' is currently an Assistant Professor of computer science at the State University of New York College at Brockport. His interests include information technology education and the World Wide Web as an information system for creation, discovery, storage and dissemination of knowledge. He has over twenty years of academic, industry, and government experience in applying information systems to solve large-scale problems. He has supervised multiple large and small software development projects as well as complex hardware/telecommunications designs and installations.

Lorraine Staehr is a Lecturer in the Department of Information Technology at La Trobe University. Her research interests are in mathematical modeling of adsorption processes, women in computing, the business benefits of ERP systems, and information systems education. She has published research papers in a number of international conferences and journals.

Leigh Stelzer's current research focus is Web-enhanced instruction and its relationship to student individual differences, learning styles, and teaching strategies. He has published professional articles on trade negotiations, joint ventures, and cross-cultural problems doing business in China. Additional research publications examine the strategies for effective business education and the interface of business and politics. Dr. Stelzer is a Board Member and Past President of Beta Gamma Sigma Alumni of Greater New York and Past Chair of the Department of Management, W. Paul Stillman School of Business, Seton Hall University. Dr. Stelzer serves as Chair, Intranet-Based Management Education of the Management Education Division (MED), Academy of Management (AOM). He received his BA from New York University and his PhD from the University of Michigan.

Arthur Tatnall is a Senior Lecturer, and Director of the Electronic Commerce Research Unit in the School of Information Systems at Victoria University in Melbourne, Australia. He holds bachelor's degrees in Science and Education and a research master's of arts in which he explored the origins of business computing education in Australian universities. His PhD involved a study in curriculum innovation in which he investigated the manner in which Visual Basic entered the curriculum of an Australian university. His research interests include technological innovation, information systems curriculum, Visual Basic programming, project management, electronic commerce, and information technology in educational management.

Connie E. Wells is Chair of the Master of Science in Information Systems program at the Walter E. Heller College of Business Administration at Roosevelt University, Chicago, IL. Dr. Wells earned her BSE degree in Math Education at Northern Illinois University. She received her MBA and PhD degrees from the University of Minnesota, majoring in Management Information Systems. In addition to teaching at Roosevelt University, she taught at Georgia State University, Reinhardt College, and Nicholls State University. Her major areas of interest in teaching and research are teamwork in systems analysis and design, and in assessment/evaluation of software and information systems.

Index

A

academic programs 229
access to technology 161
action research 174
actor-network 214
adoption of PBL 122
American Associate of University Professors 166
animation 155
asynchronous 93
audio 155

B

Blackboard® 21, 159, 162
Bloom's Taxonomy of Educational Objectives: The Cl 107
Building Teams 9
business process redesign 174

C

"certification" scheme 105
CGI 161
chat groups 167
chat space 154
class projects 3
classroom technology 21
cohort approach 31
'Colonial Echo Model' 213
competition 216
computer literacy 158
Computer Science 194
computer-based training 153
Computer-supported learning 89
conflict 15

constructive 90
constructivism 111, 130, 133, 147
Constructivist theory 33, 133
copyright 165
cost of production 161
course listserver 154
course partnership 172
creative problem solving 51
critical skills 44
cultural diversity 19
Culture 19
curriculum 194
Curriculum change 213
curriculum innovation 212
cyberprofs 156, 164

D

DAMA 246
Data modeling 43
Data Modeling Techniques 47
Day & Zimmermann, Inc. 172
deep learning approaches 60, 77
DePaul University 194
design technique 45
Didactic learning 109
"digitized text" 154
disciplines 229
diversity in teams 7
dynamic lecture 163

E

e-commerce 194
e-mail services 154
E-R diagram 45
eCollege.com 162

Copyright © 2002, Idea Group Publishing.

ecological metaphor 213
ecological niche 216, 222
ecology 215
ecosystem 216
education 106
Education for Information Resources
    Management 252
educational technology 130-34, 140, 144
electronic journals 153
electronic university 168
email overhead 167
emergent forms of education 131, 135,
    138-140
experiential learning 47
exploitation of resources 216
Extensible Markup Language 161

F

fear of the technology 158
file transfer protocol 155
FirstClass 162
flexible education 131, 135
flexible exploration 140-142
flexible learning 130, 136-137, 147
forming 5

G

Gaia hypothesis 215
global communication 80
global enrollment community 167
Global Information Issues 255
grade 16
graduate education 28
graduate research degree programs 30
group formation 114
group work 120

H

Habitat 216
History of IRM 248
HTML 155, 161
Hypertext Markup Language 155

I

Industry-university partnerships 173
Information Security & Quality 256

Information Systems 212
information systems graduates 105
instructive 90
instructor-delivered courses 153
instructor-student e-mail 154
integration of cyberlearning technologies
    153
Integration of IRM 256
intellectual property 165
intellectual property issues 165
interactive sessions 167
Internet cyber cafes 156
Internet for Dummies 161
Internet kiosks 156
Internet-based courses 154
IRMA 246
IRMA/DAMA 2000 193
IRMA/DAMA Curriculum Model 246
IS '97 193
ISCC '99 193
IT education 59
IT workers 228

J

Java 161
JavaScript 161
"Just-in-Time" learning 169

K

Knowledge Universe 164

L

lack of reasonable access to the technology
    158
Learning group formation 114
learning outcomes 62
learning resources 115
legal, ethical, and social impacts of IRM
    255
level of computer experience 158
lone scholar approach 32
lone scholar extreme 31

M

Mallard 159, 162
Management Information Systems 88

Managing Conflict 15
Managing Cultural Diversity 19
managing list serves 167
master's degree 194
MIS education 89
model curricula 193, 231
models and metaphors 213
multi-species communities 216

N

NaLER (Natural Language for Entity-
    Relationship) 54
National Education Association 166
norming 5

O

on-line testing 154

P

paradigm 27, 29, 32, 153
patentable products 165
PBL 104, 106, 109, 110, 111
PBL implementation problems 122
PBL problem 121
PBL tutor 123
PC/Internet/Intranets/E-commerce 257
performing 5
"personal attributes" 108
Personality characteristics 8
Powerpoint 162
predator-prey interactions 216
problem solving 43, 115
Professional fields 229
programmed learning paradigm 163
programming 2

Q

quantitative evaluations 74

R

Rapid Application Development 214
Real-world problems 111
relational data modeling 46
relational learning perspective 77
relational perspective 62
research degree 27, 29

research methods 37, 39
research process 29, 31, 33
research programs 35
research skills 27-31, 33
research training 28, 30-31, 33, 37
rich pictures 73
royalties 164

S

"scaffolding" 111
scholar approach 31
self-appraisal 54
self-awareness 50
self-evaluation 116
self-formed teams 7
skills 108, 248
social interactionism 132-134, 138, 147, 168
sociological determinism 131
stability of technology 163
storming 5
"strategic partnerships" 105
Student Attendance 164
student interaction 167
student learning 61
student-centered learning 106
"student-enabled learning" 109
students' perceptions 62
subject-based learning 112
Synchronous 93
systems analysis and design 2

T

"teacher-enabled learning" 109
teaching space to support PBL 119
Teaching Teamwork in Information Systems
    2
Team and Project Assessment 6
team building 6
team forming 6
team management 6
team size 6
teamwork 2, 255
technology-enabled course 158
Temple University 172
Tertiary Entrance Rank 75
Total Quality Management 69
training 106

U

undergraduate curricula 60
Understanding Human and Organizational
     Behavior 255
UNext.com 164
University of Hawaii 166

V

video streaming 154, 155
virtual classroom 154
virtual university 166

W

weak students 121
web "glitz" 155
Web-based collaboration technologies 173
Web-based courses 152-154
Web-based education 130, 131
Web-based instruction 133
Web-based technologies 176
Web-CT 159, 162
Web-enhanced course 153, 154, 159, 161,
     164
Webmentor 162
work for hire 165
work group 3
workforce 248
workshop approach 36
World Lecture Hall 154

X

XML 161